HAMMER HEAD

Author's Note
Hammer Head is a work of nonfiction. Certain names and
identifying characteristics of the people who appear in
these pages have been changed.

Copyright © 2015 by Nina MacLaughlin

For information about permission to reproduce
selections from this book, write to Permissions,
W. W. Norton & Company, Inc.,
500 Fifth Avenue, New York, NY 10110

For information about special discounts for bulk
purchases, please contact W. W. Norton Special Sales at
specialsales@wwnorton.com or 800-233-4830

Drawings by Joseph McVetty III
Manufacturing by Courier Westford
Book design by Abbate Design
Production manager: Anna Oler

ISBN 978-0-393-23913-3

W. W. Norton & Company, Inc.
500 Fifth Avenue, New York, N.Y. 10110
www.wwnorton.com

W. W. Norton & Company Ltd.
Castle House, 75/76 Wells Street, London W1T 3QT

1 2 3 4 5 6 7 8 9 0

HAMMER HEAD

The Making of a Carpenter

———

NINA MacLAUGHLIN

W. W. NORTON & COMPANY

NEW YORK | LONDON

FOR MARY

CONTENTS

HAMMER HEAD

How do we decide what's right for our own lives? A close friend posed this question to me, and it echoes often in my head. What shape do we want our lives to take, and, if we've had the fortune to figure that out, how do we go about constructing that life? In Ovid's *Metamorphoses*, the gods are the reigning agents of change, and repeatedly "give and take away the form of things." People are transformed into owls, bears, horses, newts, stones, birds, and trees. Without the gods to guide us, to cast their spells of transformation, how do we become something other than we were?

I used to be a journalist. Now I work as a carpenter. The transformation, like the renovation of a kitchen, happened first in big bashing crashes and now has slowed as it gets closer to complete. In college, I studied English and Classics and engaged in the abstractions of ancient history and literary theory. A journalism job followed, and with it, continued interaction with intangibles (the Internet, ideas, telling stories with words). The world around me, material reality—the floors and cabinets, the tables, decks, and bookcases—all of it was real enough to knock or kick, but it was an afterthought, taken for granted, obscured by the computer's glow. After nearly a decade working at a desk in front of a screen, I longed to engage with the tangible, to do work that resulted in something I could touch. I grew more interested in making a desk than sitting at one.

In the *Metamorphoses*, mortals are transformed by the gods for two reasons: to punish and to save. My shift from journalist to carpenter was neither punishment nor salvation. It was an unexpected veering, a welcome re-forming. Under the guidance of my boss Mary, a carpenter and unexpected mentor, I've been given entry into the material world. I've watched, again and again, as one thing becomes something else—the way a seed becomes a tree becomes a board becomes a bookshelf. For people, such transformations are subtler, and perhaps more difficult to achieve. We cannot take a hacksaw to our habits, after all. But as Ovid writes, "By birth we mean beginning to re-form, a thing's becoming other than it was." This book is a story, a simple one, of things becoming other than they were. It's a story, like all of them, of transformation.

Chapter 1

TAPE MEASURE

On the distance between here and there

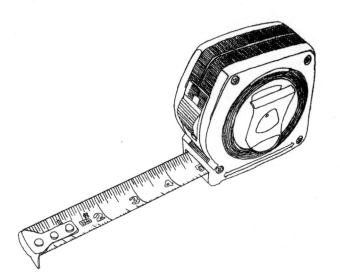

From the sidewalk on Memorial Drive where the Mass Ave Bridge begins on the Cambridge side of the Charles River, the view across extends a little less than half a mile. To the south, the Boston skyline rises above Storrow Drive. Closer to the water, and lower to the ground, brick predominates; glass and steel rise behind. To the west, moving upstream against the current, the Citgo sign lights up over Kenmore Square, and if it's a home game during Red Sox season, the floodlights over Fenway make it daytime in the park. The river bends and snakes its way out of the city, through twenty-three towns, sidewalks and river paths giving way to shoreline with pine and maples. Great blue herons stand in shallows on stalky legs and box turtles with warm shells sun themselves on rocks and logs. For eighty miles, the river wends through eastern Massachusetts from its start at Echo Lake in a town called Hopkinton. To the east of the Mass Ave Bridge, back near the city, sailboats dip and swerve. Oars on the eight-person sculls thunk in the oarlocks as crew teams run their practice and glide underneath the bridge. The Red Line train crosses over the Longfellow Bridge about a mile downstream. Beyond, the new Zakim Bridge rises above the river, suspended by white strands that look like the skeletons of wings. The river meets the harbor, freshwater merges with salt, and the Charles River is altered and absorbed into the Atlantic.

For seven years, I crossed the bridge on foot, once in the morning, sun at my left shoulder, and once in the evening, when sunsets sometimes blushed the sky. It was part of the three-mile path I made from my apartment in Cambridge to the newspaper offices where I worked in Boston. On the way home, depending on weather and time of year and if it was a deadline day, bands of pink spread across the sky upstream, or else it was cold and city dark, and lights became the thing, streetlamps, headlights, tail-lights like embers, all blinking and sparkling up the road ahead. The river glittered with Cambridge above it, squatter than Boston, lower to the ground. Sometimes, the moon. Sometimes, a few stars. The wind blew stronger on the bridge. Tourists handed me their cameras and asked me to take photographs with the river and the skyline. I dodged joggers and cyclists on the sidewalk afraid of the bike lane. I was usually alone when I walked the bridge, occasionally drunk, a few times crying, one time kissed by someone I didn't like too much. The walk across the river was a ferrying for my brain—toward a desk and noise and tip-tapping of keys, clicking and interviews and story ideas, and away from my desk in the evenings, toward quiet and home, toward a bar, toward not having to talk or think or be clever or click. Oh I am fond of that bridge, the whole stretch of it. It's the longest one to span the Charles at 2,164.8 feet. That's 659.82 meters, or 364.4 smoots.

Oliver Smoot was the shortest pledge of MIT's Lambda Chi Alpha fraternity in 1958. Late one night that year, he was tipped head over heels, over and over, across the length of the bridge, Boston to Cambridge, by his fraternity brothers. They made an official tally of 364.4 smoots, plus or minus one ear. Ever since that fabled measurement, twice each year the boys of Lambda Chi Alpha have repainted the markers on the sidewalk across the bridge that delineate every ten smoots. (An exception has been made for smoot-mark sixty-nine, which as of this year, had the addition of "heaven" painted underneath.) When the bridge was reconstructed in the 1980s, the sidewalk slabs were made in smoot-lengths, as opposed to the six-foot standard. Oliver Smoot's contribution to measurement continued well after his fraternity days. A plaque at the base of the bridge commemorates the 50th anniversary of the smoot, and notes that Ollie went on to head both the American National Standards Institute and the International Organization for Standardization.

I trotted across the bridge, face reddened by wind in winter, sweat soaking the back of my shirt in summer, and I went to my desk at a newspaper where I'd landed a job out of college. First I did listings, which meant inputting the city's every concert, contra dance, art exhibit, comedy show, poetry slam, and movie time into a massive database week-in, week-out. I wrote about cheap Salvadoran restaurants, interviewed David Copperfield, profiled an art-porn collective, reviewed documentary films, covered a conference on virginity, and wrote about books and authors and the literary scene in Boston. Eventually I got bumped up to managing editor of the website, which meant I was tasked with making sure every story showed up in the right place at the right time. It meant a lot of clicking.

For a long time, I loved it. I loved the rhythm of the thing, the peaks and lulls, the energy of a room of people, mostly men, on deadline. All the furious typing, all the opinions and shit-talk, listening to writers on the phone with sources, the concentration and filing and release—the newsroom possessed a charge. And I was proud to be part of it. What good fortune, to be able to go to a place every day and be surrounded by all these smart maniacs telling stories, all working to produce this thing that had a history, that was part of the fabric of the city, that was committed to long-form, investigative, issue-based journalism and had the strongest set of arts critics in Boston.

What a set of weirdos sat at these desks with me, what a collection of brains. There was the sharp-witted, chain-smoker with untucked shirts and rogue charm who had worked moving houses before becoming a journalist. There was the practitioner of make-the-world-better journalism and expose-injustice journalism, who sat at her desk and worked with the focus and fire of someone possessed until you got her out to the bar, where she'd talk about how she'd followed the Grateful Dead. The managing editor was a first-rate grump, a big-hearted cynic who had helped start the paper, and still believed in its power and necessity. The arts editor with the encyclopedic memory threw cursing fits, slamming books on the floor of his cube, his standards unmeetably high. And the features writer, from hard-knocks Brockton, wrote a weekly column about the city's strangest characters, which struck me as maybe the coolest job in the world. In my head, she towered tall above me; I saw her not long ago, and realizing that she and I were the same height came as an immense shock and had me questioning, for a moment, if perhaps she'd contracted some shrinking disease. Such is how these people loomed.

I couldn't believe how lucky I was. Every time I got to answer the question *what do you do?*, I felt proud to answer. This was exactly what I wanted. Until it wasn't.

Talk of *readers* turned to talk of *users*. Print was sputtering, and it was the responsibility of the web operation to inject "youth" and "relevance" into the operation to keep advertising dollars coming in, and keep the paper in business. It's a familiar story now.

The clicking started to get me down. There is a dullness in all forms of work, a "violence—to the spirit as well as to the body," as Studs Terkel put it in *Working*. There are repeated tasks and empty time and moments you wish you were swimming. These are unavoidable, even in jobs we love and feel proud to have; these are natural, even if you've found your calling. It's when those meaningless moments pile and mount, the meaningless moments that chew at your soul, that creep into the crevices of your brain and holler at you until ignoring them is not an option. Deadening moments that lead to the hard questions, the ones that swirl, in the broadest sense, around time and dying.

After years in which most of my waking hours were spent in front of a computer screen clicking buttons, I realized I'd become a lump in a chair, present only in the physical fact of my flesh at the desk, soul staling like a Saltine. It got worse by the day, like a shirt that had once felt so comfortable, so flattering and familiar, but had started to tighten, constrict at the neck, pull across the shoulders. The grooves of my brain seemed to be getting smoothed out, a slow dulling, gradual and slothing. It was harder and harder to find pleasure in the atmosphere or meaning in the endeavor. The people I liked most were starting to move on to other jobs at other places.

The screen exerts an oppressive power, and I am as seduced as anyone by the clips and pics, the news and noise of the Internet. I would rather e-mail than talk on the phone. I have pals I know only online and am grateful for those connections. But there is no other place I can think of where one can consume so much and absorb so little. The Internet has no equal in that regard. I am leery of its siren song, the way it beckons, and of my own inability to ignore its call. It's a rabbit-hole exit, a tumbling in space with Wonderland ever always one click away.

My brain went bad. Hangovers hounded me three out of five workdays a week. Mouse in my limp, damp hand, my head raw and frayed, I spent months thinking, *I've got to get out of here.* But I had a familiar routine to cling to, and health insurance, and despite it all, I felt an allegiance to the institution. And so I stayed, kept scrolling, kept clicking. Plus, what would I do next? What could I do? Inertia and fear and laziness, the three-headed dog that keeps us from leaving situations that have passed their expiration date, growled around me for months, the way Cerberus allowed souls to enter the realm of the dead, but allowed none of them to leave.

The tipping point came in the form of an online list. As a sardonic response to *Maxim*'s list of sexy women, we published a list of the 100 Unsexiest Men. A place on the list was granted not for physical repugnancy, but for poor character, bad deeds, and general unpopularity. Scandalous politicians, misogynist athletes, racist pundits, public-figure villains of all kinds. The first time around, the list was so popular it crashed the site, and thus became a must-repeat feature. Devising and executing it the first time was stupid fun—nothing to be proud of, but no big deal. When the third annual list rolled around, I found myself dispirited. More than that: sitting at my desk making sure the number on the list matched the number on the blurb about the man, I felt desperate. It was more than stupid and my brain hollered: *You will die and this is an empty way to spend the days.*

Slumped at my computer during those unsexiest days, all I could think about was leaving. I craved something away from the screen, away from the echo chamber of the Internet. I wanted something that had a little more to do with reality. But what did that mean? Our lives online are as bound in reality as making pancakes, driving to the dump, spilling a glass of wine. At my desk, though, I felt far away from an anchor, a grounding agent, satisfaction. In a vague way, I wanted to put my brain where my hands were. These impulses were question marks, shadow urges, pipe dreams. I wanted to be an Olympic speed skater, too, but that wasn't about to happen.

I had worked at the paper for nearly the whole of my twenties. Closing in on thirty, it wasn't just disenchantment with my web job. My brain stirred with change, with the idea of a wholesale altering of life as I'd been living it. I spent months in this mode, fed up, deeply bored, trying to corral enough courage to leap.

On my way to work on a bright and mild September morning, I crossed the Mass Ave Bridge. The smoot marks, paint faded, blurred below my feet, counting out the distance. I looked at the river as I rehearsed what I would say to my boss that day. I reached the Boston side of the river with resolve but mostly fear and some hands-in-the-air hope. When I got to the office, I quit.

It wasn't just the job that ended. I moved out of my apartment, broke up with a boyfriend, and left the city for a little while. Sledgehammer, slam, dust, done.

My days were blank, every day an emptiness. The fear—that I'd never find work again, that I'd made a very bad decision, that I'd derailed myself with no chance of finding another train—morphed into regret, that sick feeling of knowing that time only moves one way with no chance to change what's been done.

Small efforts and loose routines were weak antidotes. One teary morning in early spring, doing my daily click around the Craigslist jobs section, reviewing, once again, the same few posts in the Writing/Editing and Art/Media/Design sections, I clicked on the Etc. category. Amid postings looking for dog walkers, surrogate mothers (up to $40k; tempting), and catheter users ($25 for your opinion; less so), I came across a line of text that registered itself in my chest with a quick extra thump of my heart.

Carpenter's Assistant: Women strongly encouraged to apply.

This simple post seemed to glow, holding in it the promise of exactly what I'd been craving. My fingers fluttered above the keyboard, ready to write the note that would convince this person that I was the right woman for the job.

I tried to explain my experience. None. None at all. I tried to think what might qualify me. I didn't know the difference between a Phillips and a flathead screwdriver. Should I admit that? No, don't admit that. I explained that my professional background had more to do with putting together sentences than working with hammers and nails and wood, but that I was curious and hardworking, and that I longed to work with my hands. "What I lack in experience," I wrote to this anonymous poster, "I would definitely make up for in curiosity and enthusiasm."

I pressed SEND and the initial excitement and blast of optimism was extinguished by a wave of despondency and pessimism. What a joke, I scolded myself. What a ridiculous long shot. You don't get carpentry jobs based on claims of curiosity and capacity for hard work, I admonished myself. Putting together sentences? I sounded like an asshole. I imagined the person reading my email and laughing—oh, perfect, *curious* is exactly the quality I need to help build a safe set of stairs—and then discarding my note to continue the search for someone who actually knew something. I regretted how I'd approached the opportunity, and tried to put the whole missed chance out of my head.

That same morning, I applied for a fiction-editor position at an online literary operation (unpaid), and a gig writing product descriptions of adult novelties ($20 per description, seven descriptions per week). The adult novelty place got right back to me and asked if I'd please choose one product from the list they provided and write a sample description, no more than a paragraph, demonstrating an understanding of keywords.

I scrolled through the options. Smartballs silicone kegel balls. The Liberator ramp. Bound To Please nipple clamps.

The Luxe Adonis G Spot and Clit Vibe. I heard the words of my high-school Latin teacher: when your eyes are open, you'll see classical references everywhere. *Caveat emptor.* So I put my Classics degree to work. *Adonis, his beauty unsurpassed, born out of the trunk of the Myrrha tree, was so lovely that Venus herself, goddess of love, couldn't resist him.* So began my blurb. I did not mention that lovely Adonis was the offspring of an incestuous pair, that his mother was also his sister, that his father was his grandfather. I did make mention of plunging in, the way the wild boar plunges his tusk into Adonis's groin, killing him, until Venus who loved him, changes him into a flower that blooms deep blood-red, "the very color of pomegranates when that fruit is ripe and hides sweet seeds beneath its pliant rind," as Ovid tells it in his *Metamorphoses.* Seeds and pliant rinds, plunging tusks and a beauty that bewitches the goddess of love. The petals fall fast off the flower that Adonis is transformed into, unlike the great and lasting bloom that the G Spot and Clit Vibe brings.

I sent that off, closed my computer, and took a walk in the rain.

Four days after applying for the carpentry job, four days after sweeping the thought of it out of my head, I got an e-mail back from an anonymous Craigslist-generated number. It was a woman named Mary writing to say that she was contacting forty of us who'd applied for the job, out of more than three hundred responses she'd gotten in the first eighteen hours of posting her ad. ("Sign of the times," she wrote.) This was hopeful. I'd made the short list. I let that settle for a moment before I realized that forty people was still a lot of people, and I still only had enthusiasm and a work ethic as quasi-qualifications. I kept reading.

She explained a bit more about herself, about the job, and what she was looking for, straightforward as a two-by-four to the side of the head. "I'm a 43-year-old married lesbian with a 10-year-old daughter," she wrote. She'd worked for herself for a few years and before that had worked for another contractor. "I like to think of myself as a journeyman-level carpenter and a slightly better tiler." I didn't know what this meant, but I liked the sound of journeyman. It brought to mind a wandering carpenter, tools slung over her shoulder, traveling place to place, building and fixing, humming away in worn-in workpants, a smile on her face.

It got better. She described the traits she was looking for: "Common sense is the most important thing. Next is lugging crap, you must be able to!" I gripped my left bicep and felt the muscle swell as I flexed. I can lug crap, I thought. I can absolutely lug crap. I thought of moving couches and tables out of various apartments, hauling boxes and boxes of books up and down flights of stairs. "Tools, supplies, whatever," she wrote of what we'd lug. And common sense: sure, my judgment was sound enough in practical matters. I'm not the most practical-minded, but I'm a good parallel parker, I can follow a recipe, sometimes I know what I'm going to wear the day before I wear it. Skills used will vary from job to job, she explained, and jobs range from a day to several months, usually averaging about two weeks. And then came a list of the sorts of work that the jobs entailed in a language mostly unfamiliar. "Go in patch walls and paint." (Clear enough, I could paint, but who knew what patching meant?) "Put in a wood or tile floor. Add trim." (Sounded doable.) "Larger jobs: kitchen and bathroom renovations, structural work." (This sounded serious and intimidating.) "Demo, framing, insulating, fire stopping, boarding, mudding, installing windows, finish trim work, install cabs, porch rebuilds. Pretty much everything except additions and roofs." What did these words mean? Demo? I thought first of demonstrations. Framing? Framing pictures, I imagined, and that'd be cool to learn. Boarding? I pictured boarding houses and torture techniques, and figured it was neither. Mudding. Mudding? All of it sounded mysterious and appealing.

She asked that we explain a little more about ourselves and why we wanted the job. In my response, I tried to be as direct and honest as she'd been. I'm thirty years old, I wrote. I spent the past bunch of years working at a newspaper. In terms of carpentry, I wrote: "I'll be honest: I don't have much experience. That said, I'm strong (lugging crap is no problem at all)." I claimed a good sensible head on my shoulders and emphasized again how curious I was to learn this stuff. I wrote about the satisfaction of putting together a good sentence, but that something more immediate, more physical, more practical and tangible appealed to me, and had for some time.

"This is work I want to learn and do," I wrote. "You would have to teach me, but I would learn fast and don't mind doing hard work. I can start immediately."

H ow acute is your internal clock? If someone were to ask you to mark a minute without counting out the second ticks, how close would you come? And if someone asked you to mark three and seven-sixteenths inches without a rule, how close would you be? A quarter inch off? Three-quarters? How well does your brain know space?

The earliest systems of measurement were based on the body. A cubit was the distance from the crook of the elbow to the tip of the middle finger. A half-cubit, or a span, equaled the spread between thumb and pinky tip. What we now call an inch was the width of a man's thumb, or the distance from the tip of the forefinger to the first knuckle. A foot: the foot. In ancient Egypt, monuments were built based on the Sacred Cubit, the standard cubit plus an extra span. Two strides equaled a pace, or five feet in the Roman standard. A thousand paces made a mile. The line between King Henry I's thumb and nose measured a yard. Two yards made a fathom, or the distance of both arms outstretched. In the thirteenth century, King Edward I's Iron Ulna, named after the long bone in the forearm, set the measure for the standard yardstick. A foot was a third of the yard, and an inch was one thirty-sixth of it. Edward I's flamboyant son Edward II decreed it otherwise in 1324. Three round, dry barleycorns made an inch in his book. But nature is fickle, and the size of seeds, like fingers and feet, can't be counted on. (What power kings wielded, when the length of their bones—or fondness for barleycorns—could become the basis of standard measure.)

Forearms and pace lengths in the west, it was otherwise in ancient India, though the scale of measure there still found its distances in the natural world. A *yojana* measured the distance an oxcart could cover in one day. A length, we can suppose, that depended on how energetic your ox felt, how arthritic its joints were, how muddy the road it trod upon was, or even how greased were the wheels of the cart. A *krosa* measured the distance at which the lowing of a cow could be heard, a distance that depends on which way the wind blows. A finger was divided into barleycorns, barleycorns into lice, lice into nits, nits into cow's hair, into sheep's hair, into rabbit fuzz, down, down, to the grain of dust kicked up by a chariot that cannot be divided. And though a cow's low isn't fixed, a distance emerges in our minds, inexact but imaginable—that melancholy bellow over pastures, gentle, warm-eyed beasts over there on the hill.

Things changed in Napoleonic France when the meter was adopted. It turned away from the human skeleton to a different sort of scale. A meter measured one ten-millionth of the distance between the equator and the North Pole following a straight line through Paris. This proves a trickier distance to contend with. What does one ten-millionth of that distance look like? I see ice floes and rain forests, a stick that shrinks to that tiny fraction of that big distance. I see a globe on a shelf and a small hand spinning it around.

It's changed again since then. No longer a fraction of the earth's surface, the meter is the length of the path traveled by light in a vacuum during a time interval of $1/299,792,588$ of a second. I can't conjure a rain forest, or spread my fingers out in front of my face to gauge a span, or look at my forearm to know a cubit, or hear the sound of that cow in the distance, or tip Oliver Smoot end over end. My mind can't make sense of light and vacuums and that sliver span of time. A day's worth of oxcart travel is one thing. For my feeble brain, light speed and second fragments are impossible to conceive.

In the thirteenth century, the word *journey* meant the distance traveled in one day, and later came to mean a day's work. The base of the word is *jour*, the French word for day. A journeyman is someone at a stage in between apprentice and master, someone competent to do a day's work. Distance traveled, work done, this was something I could comprehend.

Two days after the carpenter's note about who she was and what she was looking for, I got another message, this time to twelve of us, asking these dozen people to pick a date to spend half a day of work with her. "Call this tryouts," she wrote. "Pay you cash for your time and treat you to coffee, too. Now that's an interview, albeit long."

I stood up out of my chair and smiled and the heat of excitement rose in my cheeks, and mingled quickly with nerves. What should I wear? Should I bring my own hammer? Should I bring my own tape measure? Did I own my own tape measure?

I forgot about Adonis.

The April morning of my audition day was rainy and raw. I walked up the carpenter's block wondering if she expected me to have a tool belt.

She lived on a short side street in Somerville's Winter Hill neighborhood. It won't be long until the old beauty parlors, takeout Thai spots, and the check-cashing depot shut down to make way for dark bars feigning vintage and boutiques selling handmade bags and local honey. A big brick church dominated the south corner of the street. Men in funeral suits, shoulders hunched under umbrellas, stood waiting for people to arrive. Across the street in a corner deli, people at the counter leaned over egg sandwiches and read the *Boston Herald*. A woman said her goodbyes to the lady behind the counter by name, raising her cup of coffee as she moved through the door. When she saw the funeral men, she bowed her head. Large vinyl-sided triple-deckers, the kind you see all over Boston, Cambridge, Somerville, lined the rest of the block. A moldering Victorian stood like an aging queen at the other end of the street, all turrets, bay windows, and spiraling trim. The carpenter's house was large and tall, the color of lemon pudding, with chocolate shutters, multifamily by the look of it. On a tarmac playground across the street, hundreds of elementary-school kids ran and screamed and shot hoops and avoided puddles before being hustled inside with the jangle of the first bell to start the day of school.

The carpenter stood at the end of the driveway across the street from the schoolyard, hands in the pockets of khaki cargo pants. I'd expected a larger woman, muscled and broad. She was a couple inches shorter than me, narrow shouldered, small framed. Her ragg wool sweater had holes in the elbows, and when she reached out her hand to shake mine, she smiled wide, revealing crooked teeth, the front two with a wide gap, the right one stained and snaggled at an angle. Her dark eyes shined kind. Her shoulders were set forward, the not-quite-hunched posture of a thirteen-year-old boy, confused and hiding new broadness, or of a woman not in the habit of throwing her shoulders back to emphasize her breasts. The gray-and-blue-striped woolen cap she wore over short coarse hair, salt and pepper, lent her an elfish quality, and her voice, when she greeted me—"So you're the journalist"—was higher than what looked like would come from her face. "Mary," she said as we shook hands. "Nice weather."

As we climbed into her white minivan, a rusting roaring tank of a vehicle, she explained that we'd be tiling a bathroom floor in a house in Cambridge. The seatless back of the van was loaded with tools for the day's work. Tool buckets, saws, a drill, sponges, levels, and trowels made chaotic piles in the back. A ripped sack of sandy gray powder slumped in the corner near the back door, and from the bag's leak, the powder piled on the floor like sand in an hourglass. Pieces of pale wood of various lengths were scattered like pick-up sticks. The front seat was a scramble of orange peels, a browning apple core, a chunky tape measure, a jar of salted nuts, water bottles, a tampon, a paintbrush with bristles crusted tough, a utility knife, and bags and bags of Drum tobacco, crumpled and mostly empty. Shreds of tobacco nested in cup holders, in the seams of seats, in the crease where dashboard met windshield.

When we arrived at the house, a stately old home not far from Harvard Square, it was clear we wouldn't be the only ones at work on this place. A stallion of a pick-up truck parked out front leaked testosterone out of the gas cap, and we shared the driveway with two other work trucks. A painters' truck had ladders strapped to the roof and drop cloths and paint cans in the back. The plumbers' truck had greasy toolboxes full of wrenches and white tubes and fragments of metal pipes.

Nerves began again to beat their electric wings in my stomach, and my mouth dried up. It was one thing to have this one
woman witness my incompetence; but a whole work crew of
masters, experts, professionals? It was like having a team of
hotshot race-car drivers in the backseat during your first time
behind the wheel.

Inside, the workers hustled and clomped. The place had
just been bought, Mary explained, by an architect named
Connie and her husband. They were due to move in six days
from now; an under-the-gun energy filled the rooms and halls
as men with tools did their work. "No way it's all getting done
in time," Mary whispered to me. Hammerbangs echoed off
the blank walls and hardwood floors and high ceilings. The
scream of a power saw came from somewhere upstairs. Men's
voices, a radio playing NPR, a thump of wood against wood
as something dropped to the floor, the blur and bang of sound
followed us as we moved room to room. These were familiar noises; I'd heard them coming from inside other people's
homes a hundred times, the type of disturbance that registers and just as soon disappears against the rest of the aural
landscape. A saw scream from somewhere down the block,
that clean teakettle cry, brings the image of blade and dust
spray. The ring of a hammer strike, gunshot sharp, rips down
the sidewalk from a second floor somewhere, and the mind—
for a second—registers an arm raised and a pounding down.
Just as fast, it's back to buses, car horns, the chatter of someone on a cellphone, the noise inside your own busy brain.
But hammerbangs sounded different inside a house—louder,
more deliberate, yes. But in its proximity, the sound made the

fact of work unavoidable—something was getting *done* here, something concrete and specific. What that concrete and specific thing was, I had no idea, but the sounds were urgent, and that I was going to be part of the chorus that day made them louder and more real than I'd ever registered before.

In the front hall, a wide staircase swept up and took a hard left upward. The kitchen, so large it could've contained most of my apartment, had the bright and welcoming feel of a summerhouse. It was crammed with fixtures and appliances—handsome dark-wood cabinets lined two of the walls; a trough-size sink was big enough to bathe several toddlers at once, and I counted not one or two ovens, but three. How could you fill all those cabinets? And what do you do with three ovens? "That one's not an oven," Mary said. "It's a refrigerator for wine." The formal living room had tall, wide French doors that opened out onto a garden area. Hedged in, it was a magic sort of yard that felt like a fort. The first yellowing daffodil blooms were still cocooned in a green-yellow husk and the forsythia bush in the corner would explode into yellow any day. Heavy traffic ribboned along the busy street out front, but the garden felt miles away from any sort of commuter pathway.

"So. Nice place," Mary said.

We went back to the van for the tools.

"Grab the tile saw." I stared into the back, eyes scanning the scatter of tools, no idea where to put my hands. "There on the left," Mary said, gesturing with her chin. "The beat-to-shit-looking one with the tile dust all over it."

I leaned in and lifted. A well-used machine, it was caked with dried tile dust, the way dried clay coats a potter's wheel.

A shallow tray slotted in underneath the blade and came loose in my hands.

"You handle some more?"

"Sure," I said, wanting to make good on my claim that I was strong.

She placed a drill bag on top of the platform of the saw, an orange canvas bag that held her drill and screws of various length, some black and dull, some shiny silver. Drill bits nestled next to a few short saw blades the size of steak knives. The smell from the bag, which rested right under my face, was metallic, that blood tang, mixed with dust; it was the gentle background scent of attics and exposed wood. The muscles in my arms flexed under the weight. I followed Mary, who carried a big orange bucket full of tools and another smaller pail with a fat yellow sponge like the ones we used to wash my father's car years ago, as well as a wide spatula-looking thing made of shiny metal and a cardboard carton like a much bigger milk box. *Lugging crap, you must be able to*, I remembered as we climbed the wide stairs up to the second floor, then a narrower, steeper set up to the third. I liked the word lug. It sounds like what it is.

The third floor was open, carpeted a light purple-gray, with slanted ceilings. It would be a play zone for the kids, Mary said. Lucky for them. Dormer windows lined the front of the room and looked out over the street. Windows on the wall opposite overlooked that garden and other pretty yards of neighbors. A tiny kitchen with a small fridge, stove, and sink was tucked into the corner by the top of the stairs. What a hideaway, what a dream world, so far from the grown-ups below.

The bathroom, L-shaped, had slanted ceilings, too, a big window opposite the door, a tub, a toilet, a sink. The sub-floor, as Mary called it, was a pale stone color with screws in it. It made the room feel half naked, as though it had forgotten to put on its pants. We laid plastic on the floor just outside the bathroom and set up the tile saw in the doorway. Boxes of large tiles made a knee-high tower to the right of the door. We had the space to ourselves. The noises from the work going on below sounded far away.

"You'll cut. I'll lay," Mary said. I'd been relieved that we weren't sharing the floor with a crew of painters giving the walls up here another coat, that the electricians had elsewhere to work their wires. It didn't last. "You'll cut" brought the same sort of nerves I'd felt approaching an automated train-ticket machine in an unfamiliar city a minute before boarding. I gave her a look that I hoped telegraphed the message *I've never cut a tile before. I've never used a tile saw.* I shrugged and said "Okay" with a tone of here-we-go.

I stood in the doorway facing the bathroom, the tile saw on the stand in front of me. Mary, on the floor underneath the window, which was splattered with rain, stretched her tape measure across the width of the room, from the corner behind the toilet to the other side of the wall where the sink was. She made a pencil mark on the floor at the center point. She shifted toward me, stretched the tape in front of the threshold. I moved left, noticing that I was in her light. My father, engaged in projects of various scope, was forever telling my two brothers and me that we were in his light. He'd huff, head bent over fishing lures or sketches for decoys he'd carve, and say with impatience, "You're in my light," as though we were blocking out the very sun. We'd all jump to a place where we weren't casting shadows and continue to clamor. I'd been trained, I realized, to notice if my body eclipsed the light of someone else's work. I hoped this would signal to Mary that I was thoughtful and had common sense, knew about the importance of light and when to get out of the way.

From down on her knees she asked me to pass the pry bar from the bucket of tools.

"This guy?" I said, lifting a metal tool out of a pouch in the bucket. Cool in my hand, about nine inches long, one end of it flared out like a fish tail, the other bent like a lazy J. It looked like a pryer to me, something to jam underneath and lift.

"That's the one." She made quick work of the threshold. A few quick jabs with the bar underneath the wood and some pressing tugs and the threshold popped up and off. It looked effortless.

"Toss me the chalk line."

I peered into the bucket as though I was leaning into a dark well, and wondered if these requests were part of the test, to see what I knew. If so, I was failing.

"Gray plastic teardrop-shaped thing with a little tag sticking out of it."

I rustled it out and gave it a gentle underhand toss across the bathroom. Mary caught it one-handed. She shook the thing, pulled out the little tag, and a chalky blue string emerged from the holster.

"Take this," she said, holding out the gray plastic part, "and pull."

I took hold and pulled as Mary held her end with the metal tag on the floor against the wall where she'd made that center-line mark.

"Now place the line down on the mark in the doorway and pull the string tight," she said. I crouched and crawled under the tile saw and placed the string down on the mark.

"Hold tight," she said. "Ready?"

"I think so."

She lifted the string at a point in between us so it formed a low hill in the middle of the room, and then she released it. The line snapped against the floor. A puff of blue chalk dust plumed, and a thin chalky line stretched across the floor. A friend's older brother used to play a game with us with a rubber band. We held out our arms and he'd snap the band against our skin, pulling further and further away with each snap, leaving red stinging streaks on our flesh like the blue line just snapped.

"That's the centerline of the room," Mary said, still down on her knees.

She ripped open a bag of sandy powder like the ripped one in the back of her van and poured some into another orange bucket. Fine dust poured from the top of the bucket as though a Cub Scout at the bottom of the bucket was trying to build a fire with wet wood.

"Hold your breath."

She filled another small bucket with water from the tub, dumped it into the powder, and used the drill with a long metal attachment, twisted at the end like some industrial eggbeater and caked in hard gray something, to mix it.

"Toothpaste," she said.

"Sorry?"

"You want the mud to be the consistency of toothpaste."

Mud! "Okay."

Once it was mixed, Mary slopped a pile on the floor. It didn't look like toothpaste I'd want to use—gloppy and dull gray, it smelled like wet paper. She spread it around with a notched trowel that left grooved tracks. I liked the quiet scritching of the metal trowel teeth against the subfloor, and how it left smooth swirls in the mud. She took a sand-colored tile, about the size of a record sleeve, and placed it just to the left of the centerline we'd chalked. She placed the second tile next to it, to the right of the line. And then the cutting began.

Mary soaked the fat sponge and squeezed the water into the tray at the base of the saw.

"What's the water for?"

"Tile saws are wet saws."

I nodded, figured that was all I needed to know. Maybe the friction of the blade through the tile caused flames.

She used her pencil and a metal triangle with a lip on one side as a straightedge to draw a dark line down the left side of the tile. She handed it to me.

It was cold in my hand and heavier than I thought it would be. "Okay," I said, in that same here-goes-nothing way. I flipped the switch on the saw. With a wet whir, the blade spun into motion, spitting a blast of cold water up into my face. A plastic catch was supposed to lower over the blade to reduce the spray of porcelain dust and water, like a fender over a bicycle wheel, but it was badly bent, attached with duct tape, and useless. The blade spewed a damp mist of water and tile dust that soaked a line between my breasts down to my bellybutton.

"Go slow," Mary said, measuring out the next tile. These were the only instructions she gave.

I lowered the tile on the flat wet surface and aimed the pencil line at the spinning blade, which wasn't shark-fanged like the saws I knew, but a surface smooth and circular, like a few CDs pressed together. I did not trust that it would cut through this hard tile in my hands.

It did. When blade touched tile, the saw's sound shifted. Wet whirring rose to a thudding roar. The blade hit the porcelain, chewing a dark line as tile dust and water sprayed. I guided the tile, hands at corners closest to my body, tried to keep it straight, shifting, moving, steady here, steady. I shifted too much and the tile jammed. The angle of the cut pinched the blade and it came to a shivering halt with a noise that meant *wrong*. I looked at Mary on the floor, my face a stricken sort of question mark. She turned toward me and without words put her hands out in front of her, miming the guide of the tile, then moved her hands back toward herself, then forward again. Reverse, said the gesture. Back, then forward. I backed it out a bit and the blade sputtered back into a spin. I smoothed the tile forward, steered it right, kept cutting.

The saw roared. But I didn't notice the noise. I didn't notice the spray or the dust or that the front of my shirt was getting soaked. All I knew was the pencil line and the corners of the tile against my fingertips and keeping that cut on the line. I realized at one point that I'd been neglecting to breathe. I pushed the tile slowly, halfway across. Time dilated. Time lasted miles. The section broke off, dripping wet with a tiny chip on the corner. I turned off the saw and handed the cut tile back to Mary, my hands wet now, and cold.

"I chipped the corner."

"Doesn't matter," Mary said. "It'll get hidden under the baseboard." The relief made me think of doing my first Q&A for the newspaper—my editor told me the questions could go in any order I liked, that the interview didn't have to follow the way I'd asked the questions in the interview itself. How literal-minded we are when new to work. How pleasing to learn that there's slack in the toil, room for error and for play.

Mary placed the tile on the floor with the cut side against the wall. She pressed it into the grooved mud. She marked another, handed it to me. I flipped the saw switch again and wiped the first splash out of my eyes.

So we went. I took too much off one. Mary placed it, eyed the gap, said "Too small," put the tile aside and marked another one. Too big and she'd hand it back. "Just a skosh more." A sloppy and uneven cut she held up for me to see. A wobbled jagged edge, nothing straight or clean about it. I cringed.

"Sorry. I lost control of that one."

For the curved line around the toilet base, she demonstrated a piano-key technique to get the arc that the tile saw wouldn't allow. She showed me how to cut slices every half inch or so along the curve so the pieces look like long teeth in a wide smile. Then tap each piece with a hammer, or extra bit of tile or whatever you have on hand, to break off the teeth and achieve the curve. Any jagged bits can be smoothed down with a file. I liked this trick. It was tidy, quick, and commonsensical. The tile pieces made such a satisfying *tinck* when they broke off.

Then she handed me a tile that had no mark on it, no dark pencil line to show me where to cut.

"Four and eleven-sixteenths," she said. I fumbled through the bucket for a tape measure and one of the flat pencils I'd seen Mary using. I repeated the number in my head. Four and eleven-sixteenths. It was as foreign a number as I'd ever heard. Ghosts from high-school math class—geometry proofs, variables in algebraic equations—streaked around my head. Four and eleven-sixteenths. It made less sense every time I repeated it, the syllables dissolving into the damp and gritty slurry of tile dust and water.

I hooked the metal lip over the edge of the sand-colored tile, and stretched the tape across it. Mary, still down on her knees, turned her back to me to press another tile into the mortar on the floor. While her back was turned, I used my thumbnail to count the tiny lines as fast as I could. One-sixteenth, two-, three-. I got to nine.

"Are you *counting*?" she asked, her back still to me.

My face flushed. This is the point where I don't get the job, I thought. This is where my carpentry career starts and ends. This is the point where she smacks her head and says, *Girl doesn't even know how to read a damn tape!* It felt like getting caught cheating.

Using a handsaw, hammering a nail, pushing a tile against a blade that spun through a shallow bath of water, these actions would require practice. I knew this from the start, reminded myself that morning on the drive, and as I pushed the tile against the blade. I couldn't expect—or be expected—to make the tools or materials do what I wanted them to do right away. The tape measure, though . . . I hadn't thought it, of all the tools, would pose the steepest challenge.

These holster tapes are ubiquitous now in tool buckets and junk drawers. I played with one that lived in a kitchen drawer when I was a kid, clipping it to a chair leg, extending it across the floor, then jiggling it so it would come whipping back across the floor to make the holster jump and flip with the momentum of its entry.

The first patent for the spring tape measure, the kind that zips back into the holster, was granted to a New Havener named Alvin J. Fellows in 1868. His main contribution to the tool was the mechanism that allowed the measurer to lock the tape at any distance. A useful improvement, being able to stop the tape in place so it doesn't come whipping back before you've had the chance to make the measure.

But the holster tape didn't catch on until the 1940s. Until then, carpenters used folding wooden rulers instead. We had one that lived on the workbench in the garage. I got my fingers pinched in its hinges and still remember the hurt.

Whether an old-fashioned folding wooden ruler or a new-fangled holster type, the problem was the same: All my lugging skills wouldn't undo this failing, this basic lack. I knew nothing about carpentry. I knew even less than I thought.

Mary got up from the floor and stood next to me. She took the tape from my hand.

"Here," she said, pointing, "what's this?"

"Two and a half."

"What's this?"

"Two and three-quarters."

"This?"

"Two and a quarter."

She moved her thumb again. Some mortar had dried and cracked on her knuckle. Her fingers were long, feminine, strong.

"Two and an eighth?"

She shook her head. "Guess again." I leaned in closer to get a better look. I could smell the cigarettes on her.

"Two and —" The lines blurred up and my brain went blank. I touched the damp spot of my shirt on my upper chest. Mary let a little more time go by.

"Show me two and four-eighths," she said, rescuing me. I pointed to it with my thumbnail.

"Now, what's this?" She moved her thumb back to the place it had been.

"Three-eighths," I said. "Two and three-eighths."

"All right!" she laughed.

She let the tape retreat back into its case. "What's twelve-sixteenths?" she asked.

That I remembered how to reduce fractions came as a surprise. "Three-fourths!" Here I was answering questions that a fourth-grader could get, with ill-earned pride. I felt like an idiot. But not because of Mary—her questioning was patient, never patronizing, as though she wanted to make me understand, not show me how much I didn't know, the mark of a good teacher.

"Three-fourths. Right. And if you remember that twelve-sixteenths is three-quarters, then you know where thirteen-sixteenths is, and eleven- and nine-." And she told me about her old boss Buzz, a grade-A perfectionist and skilled builder who wanted things exact to a thirty-second of an inch. "I counted, too," she said. "Practice," she said. "It comes with practice."

I practiced. We repeated the process that day. Concrete actions: first this, then this. Measure, mark, cut. The sounds of the saw, the spray, the cold dry tile, the wet cut tile, my body positioned in front of the saw, eyes locked on the line, everything else—time and language—evaporated in the concentration, in the physical act of doing.

The newspaper taught me what rote was. Sitting at my desk in the newsroom, fingertips tip-tapping, click-clicking, the dull glow of the screen reflecting off the pale skin of my cheeks, my glazed eyes, I'd felt a brain-dulled mechanization, action without thought, action without meaning or purpose. But here, with the tiles, each one had its place, part of the whole, each measurement had a purpose, each cut. There was no slumped semi-consciousness. It was repetitive, yes, but somehow not boring. The sense I was getting on this try-out day was that even if you cut thousands of tiles, even if you spent a year at the tile saw, you'd still have to pay attention. Every time. You'd get faster. You'd get better—straighter cuts, less jamming of the blade—but you'd still have to focus. The repetition with the tiles provoked presence, a specific physical hereness.

"Smoke break," Mary announced and headed out for a cigarette in the rain. "Don't tell my wife." I looked at the section of floor we'd completed so far. Rain hit the window and pattered on the roof above. Footsteps on the stairs and an old man appeared—he looked a hundred years old, with a

long white beard and long white hair tied back in a ponytail
that hung between his shoulder blades, like the tail of something that belonged in snows. He wore light hammer-looped,
paint-splattered pants and a white T-shirt that hung from his
shoulders like a sheet. He carried a paint can and a brush, a
dull canvas drop cloth under his arm. He set himself up on
the opposite side of the room by one of the dormer windows.

"Good to see women on the job."

I didn't know what to say. It would be clumsy to explain
that I wasn't really on the job, just trying out, had only been
on the job for a couple hours, that I didn't know how to read
a tape measure. It would be clumsy to say it was good to see a
hundred-year-old wizard on the job, too.

"Thanks. It's good to be on the job."

When Mary returned from her smoke, we kept working
without much talk and finished laying the tile. They needed
a night to set before they could be grouted, so we were done
for the day. The combination of concentration, newness, of
not knowing the rhythm of the day, made the minutes swift.
Three-to-four on a Tuesday afternoon at your desk when all
you're doing is murdering the minutes—it feels like torture
because in the back of our brains, what we know is these hours
are our only ones. They are finite and will be finished. A girl
I knew once went around to all the guests at a party and told
them, one by one, "This is your real life, you know. This is
your *real life*." What a thing to be reminded of—and how easy
to forget. I liked how the tiles looked on that floor.

We packed up the tools, reloaded the van, and I shivered a
bit on the ride back. I wondered if I'd botched too many tiles, if

my lugging had impressed, if she'd noticed the time I'd gotten out of her light.

"You freezing?" Mary asked.

"A little chilled."

She blasted the heat and the windshield wipers swept across the glass.

When we got back to her driveway I thanked her and she laughed. "Thank you," she said, and handed me seventy bucks in cash. That was ten bucks an hour and it seemed like a lot of money for what I'd done. "Go take a hot shower. Get that tile dust out of your hair." I rubbed my palm across my head. Damp and gritty, crumbs of tile dust had adhered to my hair. I thanked her again.

"Take care," she said.

These were final parting words, words you say to someone you don't know and won't see again. I headed home cold and low, a fatigue in my bones from standing all day and a recognition, in those two words, that she would hire someone else. *Take care.* I went to bed early and all the bad thoughts returned as the wind picked up and rain lashed: regret, work, money, health insurance, loneliness, missed trains, and empty calendars.

The next morning, gray but no rain, Mary called. She told me the job was mine if I wanted it. I told her that I did.

Chapter 2

———

HAMMER

On the force of the blow

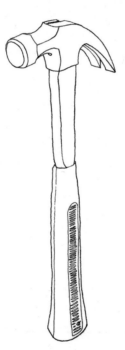

The day after Mary gave me the job, I arrived at her basement workshop. "Welcome to the wunderhovel," she said. A ping-pong table doubled as a work space; its surface was cratered and paint-splattered, pocked with nurdles of wood glue. Paint and stain cans, many rusted shut, stood stacked in towers below a small window that lit up the dust in the cobwebs that laced the paint-can handles. Power cords looped from pipes and beams above. Under mugs that held puffs of white mold on forgotten coffee, packages of sandpaper circles, old boxes of drywall screws, and an empty case for a saw I didn't know the name of, there was a workbench. Tools, rarely used, or so the layer of dust on handles and blades suggested, hung on a pegboard alongside hacksaws, long screwdrivers, and a roll of blue painter's tape. Strange clamps, with wood pressers separated by metal rods, dangled from the corner of the pegboard. A bare bulb hanging from the ceiling had a motion-detector system that flicked off the light after a moment of stillness. We stood like fools, waving our arms to make the light come back on.

Amid the mess, so much potential, so much possibility—all these tools, each with its own name and use, each with its specific strength. How powerful! We'd build walls, houses, whole worlds! Clear away the coffee mugs, spread out on the ping-pong table, take these tools in our hands, and *build*!

"You ever grouted before?" Mary asked as she rustled through a bin.

"Nope."

"Get ready to grout."

The buzz and bang of work continued at the architect's house. Mary poured dark brown grout powder from the big-milk-box container into a clean bucket. She didn't tell me to hold my breath this time, but I did anyway. She added some water and mixed.

"Grout can be a little thinner than mud," she said, pulling the mixer out of the goop and looking at it fall, the same way you'd pull egg whites up with the beater to see if they peak. She passed me a tool with a plastic handle and a flat base, like a scrub brush with a smooth white rubber base where bristles would be. Mary called it a float. What a lovely name for a tool, I thought. It conjured images of waves and small boats and surrendering my weight to water. It hooked some long-gone memory of my father taking my brother Will and me surf fishing—was part of a lure called a float? My father would fling the rod so the hook flew out over the waves and then reel as fast as he could so the bright lure whippled through the waves quick like a fish to catch the attention of the bluefish. "Some people go their whole lives without see-ing the ocean," I remember him telling us as we packed up the fishing gear on the beach one evening, putting the hooks back into his neat tackle box with the bright-colored lures with feathery, sparkly tails and such sharp hooks.

"You understand the basic concept, right?" Mary asked.

"I think so."

"Get the grout between the tiles."

We slopped grout onto the floor and began to move and spread it in the spaces between with the floats. Mary moved with fluidity. She pressed the float across the gullies between the tiles, coming at the diagonal, first one direction, then the other. I felt clumsy, trying to herd the grout into the space with the edge of my float.

"Going back and forth makes it more even," she said. I tried to adopt her technique. "You don't want any bubbles. And the more grout you leave on the tile, the more we have to clean up."

How finished the tiles looked with the spaces between them filled in, the lines even like a map of a city grid.

"I used to be able to do this without kneepads," Mary said. "Is it bothering you?"

It wasn't.

"I'm getting old." She talked of housewife's knee—I hadn't heard the phrase before. Otherwise known as housemaid's knee or prepatellar bursitis, it's a condition in which the fluid-filled sac in front of the kneecap gets inflamed, plaguing people who spend a lot of time kneeling, Cinderellas and floor scrubbers and grouters.

We finished the floor, buffed it clean with T-shirt rags, and Mary passed me the crowbar. "Take up the stair treads down to the basement."

Crowbar in hand, I stood at the top of the basement stairs. Cave-cool air rose from below with that cellar smell, damp and stony. I hoped that what I was about to do was what Mary had in mind. I jammed the bar, thicker and longer than the little pry bar Mary had used to remove the threshold, underneath the tread of the top stair. I pressed up, heaved and ho'd, and felt the board peel up and pop under my efforts with a wailing sound of nail releasing its hold on wood. I couldn't believe the force of the bar. I popped off the top of one stair, then another, ratty pieces of dark wood, faded gray and splintery where feet have landed and landed, up and down. I grunted and sweat. Halfway done, and proud of the quick work, I looked around at the basement below. A long workbench lined a far wall with an old red vise attached to the end.

It reminded me of my father's basement workshop growing up. He carved decoys down there, and the place was filled with tools. A table saw, a band saw, handsaws. Files, chisels, rasps. A mean, sharp-bladed thing that looked like a small-scale scythe. Most I didn't know the names of. With wooden-handled tools, he carved birds (piping plovers, sandpipers, wood ducks, shorebirds with long arced beaks on spindly legs). He painted them, some detailed, some crude in the folk-art tradition. He glued lifelike glass eyes into the wood. They stood on dowel legs mounted to driftwood bases he'd combed off beaches. The ducks were hollow-bodied and would float if you put them on the lake or the river to attract real ducks to shoot. The ones he made were never used in hunting. The birds he's made are beautiful, the curved shapes of their bodies and beaks, their sparkling eyes, the feathers, some dun and speckled, some green so dark it's almost black.

When I was young, I never gave much thought to the process of the making, of how a block of wood, right-angled and raw, was turned to something else. He'd go down there to the basement, and eventually emerge with a pair of plovers or a duck. They were given as gifts—weddings, birthdays. Some stayed at our house, on mantels and bookshelves. I have a small blue heron, about six inches high, unpainted, which sits on a high shelf near a window in the small apartment I share with my boyfriend Jonah in Cambridge not far from the Charles River. My father gave it to me years ago with the promise of a full-size heron someday. The only tool I was interested in down there in his workshop was a branding pencil—it heated up so hot you could burn letters into wood in dark char. I put my initials on his workbench and on scrap wood. When he and my mom split up, all his tools went into storage.

I turned back to the stairs I was in the process of dismantling and looked up to see that Connie, the owner of the house, the architect herself, forty-something in tidy clothes and an angular haircut, stood at the top of the stairs looking down at me, notebook in her hand, pencil behind her ear.

"Hi," she said, with a tone that said *Should I know you?*

I looked up at the stairs, her stairs to her basement, stripped of their treads, just the frame and dark hollows left, making descent difficult, and a flash of second-guess panic surged. Are these actually the treads? Did I just dismantle the wrong part of these stairs? Does she need to get down here? Here I was, a crowbar-wielding stranger in her home doing damage. I looked up and gave her a pained smile. "I'm just—"

"It's okay." She glanced to her right into the kitchen and something snagged her attention. "Hang on," she said to someone in there. "Whoa, hey, hang on, watch the cabinets." And she moved away from the top of the stairs, boot heels on the hardwood like bangs on a tight drum.

I took a few breaths, waited to see if she'd return. She didn't, and I continued to destroy her stairs, step by step. Mary arrived as I was nearly at the bottom, piling the cracked boards by the basement door. She looked down and nodded.

"The crowbar's amazing," I said, not letting on my doubt about the definition of tread. "I feel like a superhero."

"Good for you," Mary said. "Next time, start at the bottom and work your way up."

I looked up and realized I'd have to somehow scamper up the now treadless stairs.

On the drive over on the third day, Mary mentioned that Connie the architect had asked about me. Mary explained that I'd been a journalist and had just started on with her. The architect had said, "I thought so." I wondered how she knew.

That afternoon, Mary and I were in the master bathroom. I was perched on the side of the tub watching as Mary, crouched and bending over the shower bed, demonstrated what pitching a shower meant. She'd poured cement into the base of the shower and was smoothing it with a trowel. Thick mud, no bumps or bubbles, angled in just the right way so the water would slip drainward from all directions. She smoothed the tool over the slickness of the wet cement, a steady skim. Stroke by stroke she glossed across the surface of cement to coat the basin. It was mesmerizing. It made me think of the pleasure of watching my closest friend cook when we lived together in our mid-twenties, the way she chopped and stirred, maneuvered between countertop and stove. There is pleasure in watching someone who knows how to use tools, in witnessing skill and nonchalance with basic things. I followed Mary's movements with my eyes, spellbound.

Connie the architect appeared in the doorway.

"I found out about your secret life."

I bristled, blushed, drawn out of my trance. The sentence had a note of accusation, the unspoken charge: you are guilty of pretending. "It's not really a secret."

"Did you have a particular beat?"

"I wrote about books mostly."

She raised an eyebrow in a way that spoke surprise and approval. I was aware of the tool bucket at my feet, of the dirty jeans I was wearing for the third day in a row, of Mary crouching and smoothing cement.

"Fiction or non-?" she asked, and asked for any recent highlights. I listed off a few books, started to tell her what I'd liked about them. "There's a new debut collection of stories that's amazing—the author blends real and fantastical in this seamless way, so you're reading about this sad couple leading the sort of lives we all lead, and then Big Foot becomes a force in the story, or the Loch Ness Monster. Really lyrical and excellent and —"

"Pass me the sponge," Mary said.

I stopped, face red and heart pounding, and rummaged through the bucket. Whether Mary was trying to remind me where my attention ought to be, or just needed the sponge, I don't know. But the point was taken. I passed it to her and I went back to watching her in silence. The architect slipped away toward another set of workers in another part of the house, and I went back to perching on the edge of a tub learning how to make water move toward a drain.

S pring opened into summer, and Mary and I tumbled from job to job. We made built-in bookcases for a kitchen in Dorchester. We took down a wall, mended cabinets, and patched a ceiling in a just-bought fixer-upper in Jamaica Plain. We installed trim and painted, painted, painted for a southern belle in Cambridge. ("Oh darn, I think I want *atrium* white instead of *linen* white. Could y'all do the front hall, spare room, and parlor again?") And we redid a bathroom for a widowed grandmother who had dozens of giraffe figures decorating her small condo in Somerville. I learned piecemeal, skill by skill as the job required.

I was delighted by the variety of what we were up to, the speed at which we moved from one small job to the next. But Mary was frustrated. She bemoaned the state of the economy—no one had money for big projects, so she was forced to cobble together odd jobs and fix-it gigs instead of the larger and more lucrative renovation carpentry work she loved and was qualified for. A day's work here, a few days there, ten days, then out and on to the next thing. Here's your back deck, your new windows, your wall. Mary talked with longing of apartment overhauls, kitchens redone floor to ceiling, work that would keep us at a place for six weeks, a couple months.

During a tedious stretch of days painting, Mary and I sometimes went an hour or two without exchanging a word. It was a comfortable silence that suited both of us—I think we were both grateful not to feel the pressure of filling the time with chatter. She would roll the paint, I would cut in with the brush along baseboard edges, in corners, where ceiling met wall. The phrase *Be the work* surfaced on my thoughts, some pop-Zen saying picked up somewhere. I tried to lose myself in the spread and flow of atrium white from the paint can where it looked like a tub of vanilla milkshake, to the brush bristles, onto the wall. Wooden paintbrush handle in my hand, the paint spread creamy, thick and slick, a silken gloss that pulse-by-pulse dried to coat the wall.

Other hours we chatted nonstop.

"You know that post I put up on Craigslist? I got three hundred responses. Three hundred." This was not the first time she'd mentioned it. "Can you believe that? In less than twenty-four hours. I was getting e-mails from guys twenty years in the trades." She slopped her roller in the tray. "Sign of the times."

She talked about her daughter Maia, putting tomboyhood behind her and hanging posters of pretty boy bands on her walls.

"It sure makes time go a lot faster, having a kid."

"How so?"

"It makes you way more aware of how much time you have left."

Maia is the biological daughter of Mary's wife Emily and their good friend Henry, who lives downstairs from them with his husband in that big pale yellow two-family in Winter Hill. Four parents, one family, one roof.

"Are there other family arrangements that you know of like this?" I asked.

"Not that I know of."

"That kid must get a whole lot of love."

"I can't imagine how people do it with just two parents," she said.

All five of them used to live together in a single-family home. When they bought the place they're in now, divided up and down, Maia referred to it as the divorce.

Mary's wife Emily is a social worker with a beaming smile and tattoos of cranes and ivy on her shoulders. She teaches fitness classes and competes in triathlons. They've been together for more than twenty years, and when they talked on the phone, Mary's voice rose a pitch, there was sweetness and affection. "Hey, hon. Love you." It was a good thing to see, this tough broad with paint in her hair and a nail between her teeth being so soft and affectionate even after so many years together.

Mary is thirteen years older than me, a good distance of time—not old enough to feel like a parent, and not close enough in age to feel like a peer. Without trying, without any sort of superiority or condescension, in the natural way she has, she gave the sense *you can learn from me.* She didn't have it all figured out, nor did she claim to, and that, too, made her the ideal sort of teacher, someone who is also learning still.

But our conversation often went back to her craving for bigger jobs.

"I want something I can sink my teeth into again," she said, rolling paint on a back-bedroom wall with a bay window overlooking a narrow Cambridge side street. "All this Mickey Mouse shit"—it was her phrase for amateur stuff—"I'm losing my mind."

Not me. My mind was consistently being blown.

We spent a week building a back deck on a dead-end street in Somerville. After demoing the existing one, crumbling and rotted, we dug four post holes with a post-hole digger, a dual-sided shovel with two long handles that you plunge into the earth in a two-handed combination of throwing and stabbing. Once it's in, you pull the handles apart to bring the shovels together to grip the dirt, then lift the load out of the hole and pile it nearby. It took hours and made my shoulders ache. Each hole had to be four feet deep, which is a long way into the earth, a depth that brings coffins to mind. Four feet is regulation depth in the northeast to support a structure like a deck; that's the depth that reaches below the frost line, Mary explained. In winter, soil freezes from the surface downward. Low temps creep in and down, the way on certain days in February the cold seems to seep below the surface of your skin and deep into your blood and bones. As water in the soil turns to ice, it expands under the earth, and presses in at whatever it comes up against with magnificent force. Tens of thousands of pounds per inch can shift a fence post, a beam, a building. When, come spring, you see a fence post that's risen from the earth, uneven with its partner posts, heaving is what's happened. The freeze makes what's under the surface shift and heave, like a chest rising with a big breath in the lungs, so it's important to dig fence-post holes below that frost line. I'd known none of this, had never considered dirt and water and cold and their relationship to deck posts, all the action underneath the surface, the stuff we never see.

I started to notice stoops and decks and porches everywhere I went. I looked for the greenish tint of pressure-treated wood. It used to be impregnated with arsenic and other chemicals to repel water and prevent rot. When Mary told me about the arsenic, I started holding my breath as I chopped pieces for our deck. Pressure-treated wood is heavier than regular wood, and has a strange cold dampness to the touch. As I moved about my little world, I saw decks everywhere, with potted geraniums and hanging ferns. Twinkly Christmas lights twisted around railings. Balusters had bikes locked to them and waterproof cushions softened seats. Decks everywhere, each with wood that had been measured and cut by someone, and we were building one.

It was like standing front row at a parade of things I took for granted. Stairs, for example. Useful for moving between floors, for reaching your front door, for heading underground to catch a train to another part of the city. Codes regulate height and depth. We all know the feeling of a stair rising higher than the one before it, catching our toe on its lip; or more jarring, in the descent, stepping down with the expectation in your every bone that a solid thing will be there to meet you, to take your weight—and it *not* being there. Or it comes up too soon and sends a jolt through the ankle, up into the knee, the ugly vibration of impact. We've all felt that falling feeling right before sleep, the plunging feeling where we take a step and miss and make a fast thrash in our sheets. Muscle memory is fast formed—our bones know where the next step should come— and it's important that steps answer those expectations. Rules for stairs go back a long way.

In *De architectura*, his ten-volume, first-century-BC treatise on architecture—and astronomy and anatomy and mathematics and color ("I shall now begin to speak of purple, which exceeds all the colors that have so far been mentioned both in costliness and in the superiority of its delightful effect")—Vitruvius proposed, "The rise of steps should, I think, be limited to not more than ten nor less than nine inches; for then the ascent will not be difficult. The treads of the steps ought to be made not less than a foot and a half, and not more than two feet deep." In the eighteenth century, French architect Jacques-François Blondel suggested in *Cours d'architecture* that the length of the human pace should dictate the rise-versus-run ratio, the ratio of step height to step depth. American builders closer to our own time subscribed to a useful approximation: that the sum of the rise and the run should total about seventeen and a half inches. Now, a stair tread, where you place your foot, has to be at least nine inches deep. And the riser can be no more than eight and a quarter inches. No space between two steps can vary more than three-eighths of an inch.

Looking at the deck and the skeleton of the stairs, I was grateful that Mary would be the one to do the math to figure out rise versus run. The phrase alone raised ghosts from school geometry class, my sullen self chanting in my brain *I'll never need this* the whole way through—a rationalization for my lack of effort and ability. Mary determined the height of the rise between the steps and depth of tread, and how many steps we'd need to get from the platform at the height of the back door down to the ground. I cut planks for treads and risers and couldn't believe what was happening. Three days ago, if the man who owned the house had walked out his back door, he would've fallen out and maybe knocked his skull on a deck post. Now there was a platform and seven stairs down to the ground. We hadn't built the pyramids or the Parthenon, but this was something. When we'd fastened the final post cap to the deck, I climbed the stairs, grinning, going from ground up to the landing by the door, up those seven steps. I clomped on them, tested their strength.

"Can I jump on it?" I asked Mary.

"Knock yourself out." So I jumped hard on the platform. Solid. Nothing shook. It took my weight. Mary reached up from the ground and hopped to grip the side of the deck and did a pull-up. "Pretty sturdy."

We'd built a way to get from a door down to the ground, a passage and a place to pause, to pile groceries, to stomp snow off boots on the way inside. What a thing!

From there, we bounced to the next job and the next. Each one, over some months, aided in lifting the curtain that had obscured the physical world closest to me. Now, there were doorways, shelves, and walls. Wood, glass, plaster, paint. The awareness, this new noticing, had an intense effect on my sense of my own body and place in the world. I wasn't just my own human sack of flesh, inhabiting mental space. There were walls around me, and thresholds. There were windows that let in light and sound, traffic noise, rain; panes that showed the shifting shadows as the sun described its curve across the sky. I knew now how many pieces of wood framed those windows and doorways, how they were put together. None of this stuff had occurred to me, I'd had no occasion to consider it, and now, with every day at work, with each new task and practiced act, it was being hammered home.

It's a truth of travel that we see so much more when we are away from what we know. Removed from home, we notice: shadows and birds and sirens, the shifting color of the sky, the peak of a certain roof, the way a set of stairs descends toward a river bank. We notice: the color of squirrels scattling up a tree; the sound of chickens squawking in the road; the smell of burning garbage, of low tide, of baked bread. We're blinded by the familiar. The sirens, smells, roofs, and sky, they're all there, too, in the place you know the best. At home, awareness, open-eyedness, requires the effort of attention. The carpentry work, in these first stages, was like being in a foreign city. All this newness, it was a defamiliarizing of the most familiar: my kitchen cabinets, the doorway to my bedroom, the bathroom tiles.

When we had a lull in work, a few days between jobs, we sometimes spent the day working at Mary's house. It was in a steady state of half-done projects and mile-long to-do lists. One or another room was always in a state of renovation and change. One morning involved the demolition of her chimney.

The demo men showed up at nine fifteen.

"These guys are something else," Mary said as they pulled into her driveway.

Rain threatened and the wind was strong. The guys thumped down out of their huge dump truck. The three of them, two young guys and the boss, surveyed the chimney with heads back, looking up at the roof. It needed to go away, from the third-story roof on down into the basement. Inside, one wall needed ripping out, three others needed to be stripped to the studs, and a ceiling needed to come down. These men were here to do it.

"All right boys, head on up," said the leader, a legend in the trade, a refuse collector and disposer, a demo-god. He had curly hair and a handlebar mustache. His big belly looked hard to the touch and his fingers were thick and dirty. Scars, gashes, and scabs, some pink, others red-black, marked and cratered the skin of his calves and shins like dark dried fruit. He talked quickly, and when he laughed, often and out of nowhere, he wheezed and his eyes darted.

His two sons worked with him. Son One, in his early twenties, looked nothing like his father. He was lean to the point of scrawny, narrow of shoulder and back. His pants fell low off his non-ass. Mary mentioned the time he had finished an all-day demo job for her and then done one-armed pull-ups off the side of the truck. His large eyes were blue, and blonde curls hung over his face and beard. He looked like a folk singer or a cult leader. And when he climbed out onto the roof, sledgehammer over his shoulder, he looked like some kind of thunder god, a thinner Thor. I couldn't take my eyes off him.

Son Two, a little older, looked more of this earth, like a

street tough in all the high-school movies. Built like his father with a softer tummy, his face was smeared with dirt. Chubby cheeks made his eyes small. He wore a Scally cap and purple sweatpants. Mary relayed a story about how he had once drunk a case of Molson and accidentally shot himself in the hand.

They didn't wear masks when they worked. They didn't wear gloves. I worried for their safety. The shit they smashed and tossed, the dust that plumed, the fragments of brick and old mortar, the particles of insulation, old plaster, the rust and mold, it went into their lungs, into their blood when they were cut. I could imagine the percussion of coughing at their house at night.

Son One and Son Two walked light-footed on the shingles up above, taking turns bashing bricks with sledgehammers. They tossed debris down off the roof into the bed of the truck. Bricks flew down with comet tails of rubble. And the sound of the bricks hitting the bed of the truck, a clanging rock-and-metal thud, echoed up and down the block.

The father, the boss, spoke about other jobs he'd had while his sons worked. In the winter, like a lot of tradesmen, he fixed a plow to the front of his pick-up and cleared driveways of snow. He had seventy houses on his roster, places he'd been plowing for twenty years, a hundred bucks a driveway. That's $7,000 a storm, with an average of ten storms a winter. Big money for ten long nights. He talked of a demo job he'd had the month before, a six-family place in Cambridge. The three of them took down the inside of the place, hauled seven tons of stuff away each day for ten days. It earned them $15,000.

"None of it's as hard as fishing," he said, and talked about the year he spent in the early eighties on a fishing boat out of Scituate, a seaside town twenty-five miles south of Boston. They caught sand sharks to sell to England for fish-and-chips. The nets ran the length of thirty football fields, he explained, and were dragged three hundred twenty-five feet deep in the sea. "You never knew *what* was going to come up in those nets," he said. "Old anchors, bits of ships, eels with human teeth, and fish this big"—those leathered hands held apart three feet—"every two feet of the net. Every two feet for thirty football fields!" he repeated with a rasping laugh, eyes wide.

He talked of driving delivery trucks, delivering Pepsi, bananas. He talked about his pig, how he had to kill it that past November. It got frostbite and one hoof blew up like a basketball. He ate some of the pig and burned most of it in a bonfire at the edge of his property. Three times a year he held a giant yard sale, he explained. He'd set up tables and tables full of the treasures he collected in his years of demo work. "You never know what you're going to find!"

So far, behind walls and under floors, Mary and I had excavated some marbles, a New York license plate, newspapers from the first years of the twentieth century, green plastic soldiers, a girl's white ice skate, laces tied in a tidy bow. They seemed strange things to find inside the walls and under floorboards. A little ghost girl appears, one foot skateless, half gliding, half padding sock-footed across an icy pond. Or ghost kids sitting at the top of the stairs, letting marbles bounce down into part of the new wall to make room for another brother or sister.

In a rare lull in his torrent of talk, I asked the demo man if he had kids besides his two sons. He talked of a youngest son, "a genius, won an award from the mayor," who hit high school and something went wrong, ended up spending time in an institution. "Like I said, you never know what you're going to get." And he laughed again, but it wasn't really a laugh.

His words echoed ones I've heard from my mother. Intending to share her wisdom, she's cautioned me since I was eighteen about having children. "You never know what you're going to get," she's said, again and again over these years. The demo man said it in reference to trash and treasure behind the wall, eels with human teeth, a son in an asylum. My mother has said, "You might end up with a monster."

The boys made short work of the chimney. The whole of it, a structure that had been there for a century, was gone in less than an hour. Now it was a hollow column of space through the core of the house, as though someone had reached down its throat and pulled out its esophagus.

They moved on to the wall and the ceiling of the dining room on the first floor. Cracks and thuds, sledgehammers into walls, crowbars between plaster and stud, and parts of the house fell to the floor. A thick dust began to drift out the window and up into the sky. The thin arms of Son One reached out now and then, letting parts of the room drop to the mulch below the window, disembodied arm limbs dropping the house out the window. He braced his hips on the windowsill and leaned his body out to drop a particularly heavy bag. He looked over at me, in my direction anyway, with those blank eyes. I gave a small wave and he pulled himself back in the window without a smile. Garbage bags of plaster and lath followed, bagged separately, the lath in neat bundles, the jagged corners

of heavy plaster fragments pushing at the plastic like aliens trying to exit some weird black womb.

The pile of wood and bags grew fast outside the window.
The boys' arms, when they reached out to drop a load, were
brown with dust. The crashes got louder inside when they
reached the ceiling.

It was unsettling, the noise and the dust and parts of
the house coming out of the window. The dismantling happened fast. Time and moisture chew in a way that's too slow
to see, as powerful as sledgehammers, but so much slower.
It shouldn't be so quick to take a house apart. It should take
more than two brothers and four tools and a roll of trash bags.
The fact is, not even that much is required. The room as it
had existed, as rooms exist, with four walls and a ceiling, no
longer did. The esophagus went first, then a chamber of the
heart. Where once a wall was, now a few thick posts and two
rooms were one, kitchen bleeding into what was once the dining room. The rest of the room was all studs and hollows,
dark wood, nothing smooth, little piles of cottony gray insulation collecting in corners of the floor. A skeleton space, and it
happened so fast. Something real and lasting, undone before
noon. Such is the mutability of a room. Such is the strength of
a big hammer.

"I got hammered in Haines, Alaska," read the
T-shirts from the Hammer Museum, a small
house ninety miles north of Juneau that displays over fifteen
hundred hammers. Cigar-box hammers, medical hammers,
paving hammers, around-the-corner hammers. Hammers that

look like axes. Hammers for testing the quality of cheese. Founder Dave Pahl left Cleveland for Alaska in 1973, just out of high school, with a pioneer drive and a longing for a back-to-the-land lifestyle. He'd spent time as a kid tinkering in his grandfather's basement shop. "The man could make or fix anything," Pahl told me. But besides his own messing around in the basement, Pahl had little building experience.

He made do. In 1980, he and his wife won a five-acre home site in the state land lottery at Mosquito Lake. They built a cabin together, and lived without electricity "until I built my own hydroelectric plant," he said. A life without plugs meant a life without power tools, and Pahl learned blacksmithing and forged over a hundred different hammers for himself. But that's not what stoked his hammer passion.

A trip with his two sons to the lower forty-eight introduced him to antique shops and flea markets.

"I bought a hammer I knew I would never use—a medical hammer, the kind they bang on your knee—and that's when the collecting started."

Cruise ships float into Haines during the summer months and Pahl works as a longshoreman. He drives thirty miles from his home, ties up the boat, and waits around town to cast off at the end of the day when the tourists have herded back on the ship. In 2001, a dilapidated house in Haines went up for sale on Main Street. Pahl knew it'd be the perfect place to show off his collection and pass the time while the cruise groups walked on land. Plus, his wife Carol had recently imposed a hundred-hammer limit in their home.

"It had to do with timing," he said. "It wasn't something that I'd really planned. It just evolved."

It took some time to get the house into shape. To put in a foundation, they used hand shovels, a wheelbarrow, and a sled. During the excavation, Pahl unearthed a Tlingit warrior's pick, also called a "slave killer." He's got it on display at the museum. It's smooth and phallic shaped, pale stone. As the accompanying description in the display case notes, the pick is "believed to be around 800 years old. It would have had an elaborately carved handle and would have been used ceremonially for sacrificing one or more slaves to be buried under the cornerpost when a new longhouse was being built."

"Finding that was an omen," Pahl said. "It made me think I was on the right track."

What is it about hammers that appeals so much? "They're so simple and so diverse. For being a piece of iron stuck on the end of a stick, they're so varied," Pahl explained. The Beggar's Chicken Hammer, for example, comes from China, where it's used to crack open the shell of clay or dough that's cooked around the chicken. The Almond Tree Knocker, used to knock the trunk of the almond tree so the nuts fall onto canvas tarps, has a rubber end that looks a little like a toilet plunger without the concave plunging part.

"The stories need to be told," Pahl said. "Just to make shoes required a huge variety of types of hammers. Nowadays people can't relate to that." When he's asked about a lack of hands-on awareness these days, Pahl falters a bit. "There are benefits to this way of life," he said of raising his sons without electricity. "The world is changing," he went on. He paused, started a sentence, started another. "I don't know if I'd advocate others to do the same." Another long pause. And he went back to museum-tour mode. "If you want to talk about carpentry, probably the most important hammer is the claw hammer."

I would come to know it well.

The café in Inman Square does a brisk business in panini, pasta, and pizza. A loyal rank of regulars lines up for takeout to keep out of the way of the waitstaff as they weave and skirt in the narrow spaces between the few small tables. Mary and I were called on to build a wall to divide the kitchen—a tight squeeze of a hallway—from the area where people eat, a wall to take the place of a low refrigerator case with slices of carrot cake, bottles of Peroni, and cans of orange San Pellegrino.

The big window at the front of the café looks out on the square. A small dark bar called the Druid across the street is an ideal place for Sunday afternoon pints in winter. Up the way is a used bookstore with creaking floors, the best ice cream shop in the city, a seafood-barbecue joint, a jazz bar with a popular brunch, and a lefty-weirdo coffee shop. Methadone patients from a clinic nearby linger outside the coffee shop with wristbands and hollow eyes. The father-and-son team behind Inman Hardware runs tabs for regulars, and the old woman who worked at the convenience store used to give Swedish Fish to people she liked. It's a good neighborhood. I lived there for four years, roommates with an old, close friend, and every Monday night we'd head to the old B-Side Lounge and fall in love with the bartenders. I'd had sandwiches from this café before, and now I was here to work. This was my first wall.

Humankind's first wall? Made of cave. Or, possibly, of flesh, as Ovid has it: "We dwelled within our mother's womb until . . . nature willed that we not lie so cramped in narrow walls . . . she drew us out into the open air from our first house." After womb and cave, pelts were dried, strung up, and turned to tents. In medieval times, space for eating and sleeping went undivided. Families slept in the same room for safety and for warmth. The rise of bedrooms, of more than one big household room where families heaped together by the hearth, coincided with the rise of reading. Walls protect. They keep out (bugs, thieves, neighbors, annoying brothers, bears, wind). They keep in (heat, secrets, your family safe at night). In New England, stone walls, as old as the first colonizers, wind through fields and forests, home to chipmunks, garter snakes, dividing meadows and farmland property. A U.S. Department of Agriculture report in 1872 estimated that some 240,000 miles of stone walls crossed the contours of the New England landscape like winding spines. No official tally of mileage exists right now; it's estimated that half of all those miles of stone remain. These walls impart an austerity to the landscape. They signal the presence and stone-by-stone effort of humans before. Each rock was hefted by hand or lugged by oxen and placed one by one, to clear fields and mark land, to pen beasts or rim a family graveyard.

Walking paths through woods, forests thick with pine and oak and birch, sunlight speckling the path between the lace of needles and leaves, I've come across stone walls through the trees, away from the path, miles from the road. There's something haunted in them. Long-gone farmers deposited these rocks here, held them, placed them, and in that effort, in the solid thing that remains, their human presence is felt, and their goneness. These walls serve as a chain backward through time. Dividing space—whether it's cow pens or countries—is a powerful thing.

Walls speak to an emotional need as much as a structural one. They protect us from wind and rain and strangers. They protect our private acts and parts. They protect us from our shortcomings and our fears. A wall broadcasts: I am vulnerable.

The café was closed for business while Mary and I worked, which gave some urgency to the job; they needed to get back to pressing panini. After we moved the refrigerator out of the way, we attached two-by-fours to the ceiling and, parallel to them, to the floor. On the right and left edges we attached two more boards that ran vertically between the floor and ceiling planks to form a rectangle. We measured and marked for the studs, the upright boards that form the frame of the wall and support plaster, sheets of drywall, or plywood.

Hanging shelves in a kitchen at another job, I'd watched as Mary knocked on the wall with her knuckles.

"I'm trying to find the stud," she'd said. "The drywall won't support the shelf. You want to make sure you're hitting wood." Another way to do it is drill holes in the wall until you hit something, feel the resistance against the bit when it hits the wood behind the wall. This technique only works if something will cover that section of Swiss-cheesed wall. Mary rapped along the wall—hollow knock-knocking, and then a duller thud. "Hear that?" She knocked again. "Hear how it's not so echoey? That's the stud." She made a mark with her pencil on the wall and placed the shelf bracket over it to screw into the wood behind the drywall. She stretched her tape, knocked again at around sixteen inches, and heard the same dull thud. X-ray eyes, I thought. "Sixteen on center," she said. "Typically you're going to find the studs every sixteen inches. There's a million reasons why it might not work out that way, but that's the rule." You can buy stud finders that beep and light up, or you can knock and listen.

We marked for the studs for the café wall, the center of each one sixteen inches from the center of the last. I held a board up straight as Mary nailed it into the plank on the floor and then the ceiling, driving the three-inch nail in at a diagonal—toenailing is the term for joining vertical and horizontal boards with nails driven obliquely. She nailed three on each side of the stud on the bottom, and three on each side at the top, twelve nails for each stud to make it rigid and secure.

She drove her nails with power and accuracy. Five strong hits, or three, and the nail was in. This is basic, I thought. I have a strong arm, I'd used a hammer before. How hard could it be?

Mary passed me her hammer, the blue rubber handle still warm from her grip. She started chatting with the two women who owned the place.

"If you ever want some catering shifts, let us know," one of them said to Mary.

"For old times' sake," Mary laughed.

"You worked here?" I asked.

"Back in the day. What, ten years ago now? That's how I got so anal about how long proteins can stay out of the fridge."

The lunches Mary brought with her to work weren't rush-packed tuna sandwiches with bags of chips. Great savory smells rose from her Tupperwares—sausage and white beans in garlic and tomato sauce, ribs she'd grilled the night before. She was always talking about pork. She took care with food and liked to eat well.

As they chatted, I gripped the hammer. With my left hand, I pinched the shining three-inch nail, stared hard at its small head, and positioned it on the two-by-four. I aimed the nail at an angle as I'd seen Mary do, so that it would drive through the vertical stud and down into the horizontal plank on the floor. The noise of their talk dissolved behind my concentration. I tried to press the nail tip into the wood to give myself a head start, to find some purchase in its hole before I took a whack. It shifted and I repositioned it, held it tight between my thumb and forefinger.

I raised the hammer and struck down. The nail blasted off, skidded and clinked across the floor.

I grabbed another from the box and tried again. Some purchase, some press of metal into wood. Quick victory. I whacked again. The nail bent to the left. I directed force in the opposite direction to get it to straighten. I swung three more times, *bang, bang, bang*. It bent more, curving under the impact. I missed the nail fully once.

"A swing and a miss," Mary said.

This is a disaster, I thought. I used the claw of the hammer to extract it, mangled, from the wood.

Another try, and this time, *bang bang bang*, it took eight hits, but the metal moved through wood and fastened the two pieces together. My heart pounded from the effort. One down, eleven to go.

What a villain a nail can be. It took on an intelligence, a sinister character—a worm, a non-cooperative enemy. Whacked wrong, the metal seems to alter form, from something strong and firm to something flimsy, crushable, and twisting. An ugly weak thing, a bent nail. But then frustration shifted back to where it belonged: the nail's not the one with intelligence. My arm and aim became the enemy, my own unskilled self.

I kept going. The muscle zone above my elbow burned with the effort. A dime-sized blister bloomed on the soft meat of my thumb-palm.

"I suck at this."

"You don't suck," Mary said. "You just need to do it hundreds and hundreds of times."

"If the blows be violent at first, the nail will be bent or sent astray, as this time it derives very little support from the wood into which it is being hammered." So begins instructions on how to drive a nail from a woodworking handbook written in 1866. Practice, patience, power, and even then, no guarantee of success. "Sometimes the greatest care will fail to ensure the straight driving of a nail."

I watched as Mary hammered. She held the hammer lower than I did; I lowered my grip. Her wind-up came from the shoulder instead of the elbow, where I had been pounding from; I altered my swing. Her hits started gentler and built force; I switched from full power straightaway and built strength with each swing.

I counted Mary's strokes. I counted mine. *Bang, bang, bang.*

Her nail was in and she was on to the next. Double those bangs, add stutters—*ba-bang*, add coaxing language (*come on now, no bending, glide right in there, pal*), and such was the sound of my own hammering.

Mary is a small woman. I have two inches and probably twenty pounds on her. Maybe twenty-five. Her wrists are slender, her shoulders narrow. When she started smoking again, and got a big dog named Red that she walked every morning, her pants started to slip off her waist. She used an extension cord as a belt one day. Petite would be a word for her if she didn't carry herself with the force and presence of someone much larger, if she weren't able to hoist eighty-pound bags of cement onto her shoulder as though she were lifting a sack of pine needles. Though she does have the girliest sneeze of anyone I've ever met—a squeaking *atchoo* that makes me smile every time it happens.

When we finished framing the wall in the café early that afternoon, it looked like a wooden cage you could walk through. Once the studs were in place, hammered and nailed, we screwed sheets of drywall to them. We covered the screw holes and the seams between the drywall sheets with mesh tape and Mary mudded over it with drywall compound that actually did look like toothpaste, pale and thick. Finish trim work included: baseboard and base cap (the decorative ridge or curved piece that sits atop the baseboard and looks like it's part of the same piece of wood), and crown molding, uniting wall with ceiling. A couple coats of paint after that, and then, something solid and lasting: a new room.

I couldn't believe it. When we broke for lunch on the second day, eating food from the café, I gushed. First there was no wall, and now there is, I said, like some stoned teenager, baffled and amazed by the truth of something basic. It seems like magic, but it's so simple. This is what all these rooms are made up of? I can't believe it!

"You could build a house," I said to Mary.

"I've never framed an exterior wall."

"Is it that different?"

"Not really."

"Do you ever think about it?"

Mary twisted a forkful of pasta. "I think about going to Alaska." She talked of taking her dog and living in the wilderness. "I could do without all the people."

After work that afternoon, I walked down my old street in the neighborhood, indulging the urge to see how it felt to walk past my old place without a key in my pocket that would open the front door. It hadn't changed, and fond memories flowed on a strong current. I felt full of myself, too. Walking past a former neighbor's house, I thought of the guy who lived there, a corn-fed blondie who rollerbladed everywhere in too-tight khaki pants. My roommate and I would run into him in the neighborhood and he'd say things like, "I'm always seeing you guys coming in or out of bars," in which he could not disguise his judgment. *He doesn't know how to build a wall*, I thought, passing by his apartment, with a self-satisfied pat on my back and toot of my horn despite bending half a dozen nails beyond recognition that day.

A few houses down, another old neighbor came out the door, a tall, bearded middle-aged dad. I remembered seeing him in tears on the sidewalk one summer afternoon a few years ago, a leash in his hand, the day the family's two-year-old golden retriever had died. "Her heart just stopped," he said, sniffling.

He recognized me as I walked by. "Long time no see," he said with a wave. "How's the newspaper biz?"

My self-satisfaction was blown away like a pile of sawdust in a wind. I fumbled and stuttered. "Oh, you know, I actually left my job at the paper. I'm still freelancing, but I'm working as an assistant to a carpenter, and we were just working around the corner, building this wall over at the café, and so it's sort of this new life and——" Blood pressed against the skin of my cheeks as I somersaulted through a hands-in-the-air explanation, as though I didn't quite believe what I was saying myself. I could sense his amusement.

"Well that's pretty cool. Where's your tool belt?"

His wife came out then, too, pretty in a northern California way, no makeup, thick hair, smooth skin, athletic sandals. She had a voice that reminded me of quilts.

"Our old neighbor is banging nails for a living now," he told her.

I laughed nervously. "Well, sort of."

We talked on the sidewalk by a ginkgo tree for a few more minutes before I excused myself. I continued down my old road, past the apartment building that looked like a ship, past the small playground, past the house that always had a bunch of bikes tangled in a heap by the porch, past my old apartment where our landlady had planted some flowers in the mulchy space next to the stoop. I'd left the café thinking, *Look at that, we did that!* But walking down my old street, I was reminded again of what I was and wasn't. It felt like a nervous charade. Hearing myself talk to my neighbor, I sounded unconvinced of any of it, even to myself.

So I went back to peek inside the café, to remind myself that the wall was real—and that we had built it. It was still standing. I wanted to go in and tap on it, to give it a little kick. Building it was steadying. The sense of permanence, strength, and control it gave was unexpected and welcome, especially against the shifting and question marks taking place in my own life. There was space; we divided it.

I looked at the café's website not long after we'd finished up. They'd posted photographs of the progress of the work. People left comments. "Better before the wall." "I understand why they did it, but I wish they hadn't." "Food's the same, who cares about the wall?"

There were other walls to care about. A few weeks later, a job took us to a big house in Brookline, a rich suburb west of Boston. The couple who owned the place were Russian and had a young son. I didn't meet the husband, but the wife was thin in a nervous way, and their son had a grayish pallor. And though their house was large, the rooms were almost empty: a couch and a table in one room, a lone chair in another in what might've been a dining room. Our voices and hammerbangs echoed off the floors. We were there to repair a rotting bay window at the back of the house.

I stood on the backyard grass and watched as Mary, up on a ladder, about fifteen feet off the ground, pried pieces of shingle and molding off the house with a big blue crowbar. In these first months I spent a lot of time this way, watching Mary work. I fetched, chopped, lugged, and watched. And there was always cleaning up to do. Despite the mess in Mary's basement workshop and the chaotic state of her van, she was a relentless tidier of her jobsites. We spent half an hour or more at the end of every day, after the last cut had been made, the last nail hammered in, sweeping, vacuuming, organizing, loading tools, making neat stacks of wood if we were returning the next day, leaving things cleaner than when we started if we weren't.

At the Russians', I stood by as Mary worked her way around the rim of the window, exposing what was underneath. Two-by-fours rose along the side of the window, framing it, running from the window base into the header, the heavy beam that extended across the rough opening. Mary's lean forearms flexed as she pried.

She knocked on the header with the crowbar and glanced over her shoulder back at me.

"This keeps the weight of the wall from resting on the actual window frame."

I collected the pieces of house that Mary tossed down on the grass. The cavity around the window looked like a wound.

At the lower left corner she paused and shook her head.

"Not good."

"What's up?" I said.

"This is not good."

I did not like the way the house looked gouged, and Mary's voice spoke something ominous.

"Bugs."

Behind the paint and drywall, wood is rotting. Slowly, maybe. Bugs gnaw at the beams that hold up a house, moisture gets in, fungi make feasts, softening the cellulose and lignin of the wood. The skeleton of a room unsettles because we cannot peel back our own skin. Time and moisture stalk. We're all of us decaying, every moment less of what we were before. We can't pry open a section of ourselves to look for leaks and rot. Seeing what's behind a wall proves a stark and immediate reminder of this fact. Doctors opened up my uncle, diagnosed with lung cancer, to remove half of one lung. When they peeled back his flesh and looked inside, they found the cancer had spread in and around both lungs: *inoperable*. So they sewed him closed again—nothing to be done. Epicurus wrote: "It is possible to provide security against other ills, but as far as death is concerned, we men live in a city without walls." You can build a coffin. You cannot build a wall against death.

Carpenter ants had made their meal of part of the window frame. I couldn't see how bad it was, but I could see the wound in this woman's house, and Mary stood on the ladder, shaking her head.

"It's pulp," she said. She grabbed a fistful and let it drop to the ground like wet snow.

I looked at the hole around the window and thought, *What have we done? Let's patch it up, seal it closed, and run away. How will we ever fix this before nightfall? How will we close this up so that at night raccoons don't climb in and kidnap that grayish boy, or wolves, or spiders? What if it rains?*

Mary shouted measurements and I cut pieces of two-by-fours

for her to slot in, to support the wood that was already there and to replace the part of the frame that had been gnawed away. I jogged between the backyard and the driveway where the saws were set up on their stands. Sawdust spewed and dusted down onto the pavement, resting in craters in the cement, and the smell of pine moved with it, bright and clean, the smell of Christmas, renewal. The miter saw screamed through the wood, and I hoped the little boy wasn't napping. Mary, on the ladder, leaned into the hole she'd made and sprayed a heavy toxin to annihilate the wood chewers. I held my breath and hoped she had, too.

The words for measurement were fluid, when we weren't talking exact numbers. *Take a blade off this*, she'd say, passing me a piece of two-by-four. The kerf of the miter-saw blade—the width of the groove made while cutting—is an eighth of an inch. *Half a blade* means a sixteenth, but it's an eyeball job: leave the tape clipped to your pants. *Less than half a blade* means almost sanding as opposed to slicing, using just a fraction of the teeth to chew off the wood. *Skosh* is the measurement she used most of all. *Just a skosh more*, she'd say. I usually took that to mean not quite a full blade, but more than half. When Mary wanted the scantest part removed, she narrowed her eyes and held up her thumb and index finger so that almost no light got through the space between and she'd say, "a millisecond, take a millisecond off this." I loved it when she talked about distance in terms of time. A millisecond meant barely anything at all because you can't see a second, or that's what I took it to mean. Builders use the phrase *cunt hair*, or CH, as an unofficial term of measure. "Take a red CH off that board." It's a thirty-second of an inch. Mary did not use the phrase *cunt hair*.

When the Russian woman came out to the back porch with her son to get a look at our progress, Mary suggested she be careful of the wasps' nest in the gutter above their heads. A volley of Slavic syllables and the woman hurried her son back into the kitchen.

"How are we going to get this done?" I asked over lunch.

"Like we always do. One piece at a time."

I still didn't buy it, and had visions of creatures moving through the wall during the night.

Watching Mary work, I tried to file everything I was learning into a cabinet in my brain. I found myself experiencing bolts of superiority. Walking down Mass Ave in Harvard Square, in the cereal aisle of the grocery store, sizing up passersby, I'd think: *I bet he doesn't know how to dismantle a window frame; I bet she doesn't know that kitchens and bathrooms require drywall that's green and more resistant to moisture and weighs more than the regular kind.*

In "The Student," a short story by Anton Chekhov, a young man walks through the woods on a grim cold spring evening, discouraged and pessimistic. He muses that "the same leaky thatched roofs, ignorance and anguish, the same surrounding emptiness and darkness, the sense of oppression—all these horrors had been, and were, and would be, and when another thousand years had passed, life would be no better. And he did not want to go home."

He stops at the home of two widows, a mother and a daughter, to warm himself by their fire. It's Good Friday, and he gives them a summary of the Gospels, the moment when Peter betrays Jesus. The older widow weeps; the younger looks as though she's trying "to suppress extreme pain." The student leaves the women and it occurs to him: if these women were so moved, "something that had taken place nineteen centuries ago had a relation to the present—to both women, and probably to this desolate village, to himself, to all people." And what joy he feels. "The past, he thought, is connected with the present in an unbroken chain of events flowing one out of the other. And it seemed to him that he had just seen both ends of the chain: he touched one end, and the other moved." An "unknown, mysterious happiness" overtakes him.

The horrors don't disappear (a thousand years from now there will be ignorance and anguish and leaky roofs), but the despair in what connects us shifts to joy. What the student feels, I think, is a simultaneous presence—a total being there—and a dissolving into something so much larger than his own self.

There were moments, pounding a nail with a hammer through wood, when body synched with task, when I became palm and hammer handle and the motion in my shoulder and my elbow, and the only thing was the movement, *bang bang*, and the connection of the hammer head and the nail head, *bang*, and the sliding of the metal through the wood. Like the student, fully present and also dissolved into something beyond myself, into the history of hammerbangs.

When I dissolved into the motion, walls vanished, all the divides and barriers. An echo sounded, a big bang that reverberates forward and back. We're all of us less of what we were each second, just the way roofs will leak ten centuries from now. We're all of us inoperable at some point. And when the walls lift, when we are linked with what came before through the simplicity of swinging a tool through space, or sharing a story, we escape for a moment the prospect of facing the great wall of indifference. And instead of fear, grave dread, despair, it's possible to find calm, and joy.

It didn't happen every time I held a hammer. Often it was just bent nails and bruised fingers. Most times it was work. But when it was right, the experience was a tapping in with the motion before and after me, and the threads that connect us started to glow. A different sort of door is opened, one that, for glimmering moments, gives access to immortality.

When I walked down my old street, or along the cereal aisle, and thought myself exceptional for knowing what sixteen-on-center means, I should've known better. It's not that I knew more than anyone else, it's that I knew something that so many other people have known and know and will know.

By four-thirty in the afternoon at the Russians' house, we finished framing the window. The shingles were back on. The wound was closed, the bugs were poisoned. It was sealed and mended, the rot eradicated and replaced with fresh strong wood. Solid and stable, inside and outside were back to being divided the way they should be. This verged on miraculous— that this was possible to do in a day. I stood below Mary up on the ladder and raised my hands.

"I can't believe this!"

Mary laughed.

Annie Dillard in a poem writes:

> *That there should be mahogany, real, in the world,*
> *instead of no mahogany, rings in his mind*
> *like a gong—*

I know that gong. It rang in my mind that afternoon. That this should be possible, real in the world—what a simple thing. It wasn't miraculous, was it? Prying a house apart, removing the rot, chopping pieces of pine, and making the wall strong again. This was a matter of knowing and tools. Everyday this happens. But the fact is, the truth is, it was done instead of not done. Instead of a hole in the air, there was a wall. Dillard locates the recognition of the commonplace, a firm and welcoming embrace of what's solid, ordinary, all around us. "Reality rounds his mind like rings in a tree," she writes. We find the real: in rings that mark the years, in gongs that echo, in the framing of a window, the solid stuff of everyday. In love, too. Of all humans, that you exist instead of not, that I've found you. It's not a miracle, is it? It's the total lack of abstraction, wholly actual. And maybe it's closer to a moment of grace, a noticing that takes on the weight of ceremony, and connects us to the world.

We packed the van that afternoon, loaded the saws and the ladder and lumber. On the drive home, Mary talked about the wasps and how bees in the winter stay warm by huddling in their hive and vibrating against each other. She said: "Isn't that incredible?"

Chapter 3

—

SCREWDRIVER

On screwing and screwing up

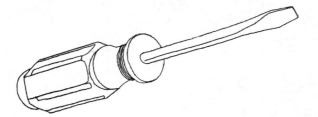

M onths accumulated. Experience accumulated. Our second fall together—shortened days and dropping temperatures—took us to a small deck job in a Somerville neighborhood near I-93. The street was densely packed with triple-deckers. The old man who ran the auto-body shop at the end of the block sat on a folding chair outside his garage, watching cars. Dust had collected on the shoulders of the tuxedos in the window of the formal-wear-rental store nearby. And I never saw any customers head into the scuba-diving shop on the corner. Car bumpers in driveways along the street wore Brazilian flag stickers.

The building we were working on signaled change to come: all clean concrete lines and well-appointed roof decks. It looked as if torn from a photo feature in an architectural design magazine. And it stuck out like a hammerbanged thumb from the rest of the neighborhood.

Four doorways along the side of the building led into high-ceilinged, track-lit spaces, each with a small front porch, a stoop with a few steps. One of these little porches had been wrecked. A resident had driven into it, and given how short and narrow the driveway was, it's a mystery as to how the person could've demolished the deck, and the car, according to a chatty neighbor, so thoroughly.

Mary and I were bundled in hats and wool socks and vests. The morning was cool and we talked about the seasonal signpost of seeing your breath for the first time, as we had that morning. First we removed what was left of the old decking, pried and unscrewed with crowbars and ratchets. We heaped the old wood to the side. The day was dry and bright. The sky, the deep blue that comes in fall, pulled everything into sharper focus. The orange extension cord snaked bright along the top of a hedge and over to our saws. A Calderesque mobile dangled high and swayed gently in someone's kitchen window, red shapes, industrial and delicate both. Seagulls landed on the roof and flew away again. We made quick work of the framing: four long joists across, two steps down, easy math. We secured the joists to the outer frame, hammered galvanized nails into the braces, and fastened it all together with long thick lag bolts. We cranked and cranked on the ratchet to make them tight.

Mary and I were in synch that morning, anticipating each other's moves, our hammer swings strong and on target, each of us focused on work with few words exchanged. This was a new pleasure, something we'd now achieve on certain days, a rhythm and connection to each other and the work. We didn't react so much as sense, as though riding the swells and dips on a river, an intimate flow. The sound of the pounding of the nails rang out, we tightened down the bolts, mirror images of each other, as the sun moved across the sky and slowly warmed the morning. We pulled off layers of clothes as the work warmed us. Language was an afterthought to our attention, almost inviolate, on the movements of the body and the action of the tools.

For the rebuilt deck, we were using Brazilian walnut the rich red color of freckles. When I sliced through it with the saw, it smelled like cinnamon, molasses, a little bit like chocolate. Other names for Brazilian walnut are embuya, imbuia, canela-imbuia; somehow those words sound the way this wood smells. Ipe (pronounced ee-pay or eye-pay depending on whom you ask) is another name for it. It's called ironwood as well, for good reason. Brazilian walnut sinks in water. To lift a board is to feel immediately that this is not like the weight of wood we're most familiar with. Feather-light cedar is twenty-two pounds per cubic foot. The mighty oak's density is forty-three pounds per cubic foot. Ipe's is sixty-six. It's so dense we had to use a special sharp drill bit to predrill a hole for every screw used to attach the deck planks to the frame. We paused when the shank of the bit got too hot drilling through the ironwood. Tiny twists of smoke rose from the holes with a sweet smell like marshmallows, and something acrid behind it, a sharp and unfamiliar scent, nothing like the homey smell of brushfires in the yard or chimney smoke; the smell itself signaled the fight the wood put up against heat. We blew on the bit to cool it off, waved the drill through the air to lower the temperature of the metal. We'd already broken one bit. It got too hot and cracked off in the hole. When I removed the remaining piece of bit from the drill, it fell and landed on my forearm, bare skin with sleeves rolled up. It left a drill-bit brand, a red burn mark painful for the rest of the day.

At lunch, we talked about the wood, its lifespan, how resistant it is to wet and bugs. Mary lamented all the synthetic decks she was starting to see. The synthetic wood can be cut like regular boards, and sprays plastic instead of sawdust.

"I get it, but who cares if it lasts until the world ends. I'm a carpenter, not a plastic worker."

We were sitting on the ground by the deck we were building. We ate lunch early. Mary's day started at four-thirty in the morning, sometimes earlier, and she never ate breakfast—just a large Dunkin' Donuts coffee with cream and extra sugar. Every morning she took her big dog to the Fells, a 3,400-acre reservation one town over. That time of year, it was dark for her walk in the woods. She came home, answered e-mails, did errands for the day, and then started to work. We took lunch around eleven-thirty, sometimes earlier. The length of her mornings, her lack of food, and her ability to function astounded me; I'm not out of bed for five minutes before I'm eating breakfast.

"And can you imagine breathing in plastic particles all day?" she said. "You'd hate it!" Mary teased me for my nerves around the stuff we breathed in, but it was comforting that she understood my fears.

"What don't you like about working with it?"

She looked at me like I had a foot for a head. "Just—"

She didn't finish the sentence, as though obviousness negated the need for words. Though Mary didn't articulate it, I think she would've said that it's like using margarine instead of butter—something chemical and false. Something that lacks soul, lacks an essence. "Wood alternative" has the same processed connotations as polyester and substitute sweeteners.

Real wood poses hassles: weather has its way with it, snow

and rain and sun break it down. It rots. Mold and mildew grow and spread. Bugs make a meal of it. Slivers of it lodge themselves in the sole of your foot or the tender skin of your palm. Synthetic wood—a composite of plastics and wood product like sawdust and pulp—requires less maintenance than real wood. Though not impervious to the elements, synthetic wood need not be treated or stained or sanded. Wood chewers like termites don't make it their meal. Synthetic wood does not give you splinters. It's usually more expensive than real wood at the outset, but possibly less over time because you can ignore it. Try as the manufacturers might, in the laboratories and factories, they haven't yet succeeded in making fake wood look real. Like a faux-fur coat with a leopard print, the grain in the synthetic wood is a close approximation of what exists in nature, but an approximation is all.

How connected can you feel to something developed in a lab? Is it possible to love something you don't have to care about? The comforting thing about wood, with its swirling grains, its knots and imperfections, its splinters and its vulnerability, is that we know exactly where it comes from. First there was dirt, a seed, sunlight, and water. Then a tree! A product of nature, and from that tree now there is a wood plank hewn from its trunk. What is polyvinyl chloride or polyethylene or polypropylene? Some people can answer that. But all of us know what a tree is. I also understand they're a dwindling resource, and I wonder if and when composite wood will come to replace what comes from forests.

In *Mythologies*, Roland Barthes laments the disappearance of wooden toys for ones made of "a graceless material" which "destroys all the pleasure, the sweetness, the humanity of touch." Wood is "a familiar and poetic substance, which does not sever the child from close contact with the tree, the table, the floor." He's talking about kid things, but the argument is the same. A synthetic deck, though easy to maintain, severs our contact with the essential. Run your fingers over a piece of raw wood, a mixing spoon, a banister, and you can sense the vibration of the natural, the warmth of the known, a subtle hum that says *this is of the earth*. Lay your palm on a deck made of PVC; there is no murmuring there, no link to forest shade or pine sap.

To witness the decay of wood on a fence that lines a field, on a forest path's fallen trunk; to see it changing color, from rich brown-red, paling to gray to green, darkening to black; to see it changing texture, from solid and strong to flaking, chewed by bugs, softened by water, dissolved by time and moisture to a pulpy mess—it comforts us somehow, echoes our own wasting, our own softening and weakening over time. There is no existential comfort to be found in artificial wood, unchanged by time, none of the melancholy that paves our understanding and embrace of time and dying. It's not that it taunts us with comparative immortality. It speaks nothing.

The Brazilian walnut was speaking. As I chopped boards for the steps and the platform of the deck, it spoke. Of weight, of toughness, of time. I basked in the day and the work, the clear sky, and the strength in my arms from months of lifting saws and holding cabinets against walls and hammering. This is so good, I thought: to be outside, to be making this thing you could stand on. The smell of the wood, sweet and charry, reminded me of s'mores. It dusted the skin of my arm and the sleeve of my shirt and I could smell it there, too. In his *Natural History*, Pliny writes that each kind of tree is "immutably consecrated" to its own specific divinity: the myrtle to Aphrodite, the poplar to Hercules. Pliny says the beech was Zeus's tree; other sources link him with the oak. This wood had the sacred about it, so dense it sinks.

In the midst of this reverie of trees, I moved a little too fast. The miter-saw blade was still spinning when I realized I'd sliced too much off our final full-length plank. It was the last long piece and would run horizontal across the front face of the deck just below the platform. Such a simple mistake—I didn't take into account the three-quarters of an inch that the top stair had added to the length of the deck. Blood heated my face. I swore under my breath. *You've got to be fucking kidding me.*

"Mary?"

"Don't tell me."

"I can—" but I didn't know what I could do. Mary had told me once how they used to rib the young guys on the crew by telling them to go grab the board stretcher from the back of the truck. Where was the board stretcher now? I told her what happened.

Mary went to the van and got her pouch of tobacco. She rolled a cigarette and smoked it and looked at the deck. I stayed quiet, my mind blank while Mary's cooked. In these moments I felt most helpless, shut out of Mary's thinking, unable to solve—or even think of solving—the problem myself. I was becoming aware of how deeply I relied on Mary to solve problems, have answers, tell me what to do. It was comfortable, in some ways, not having to be responsible for the mental heavy lifting, for the planning or problem solving. It was like riding shotgun on a long drive with a driver you trust—all you have to do is look at the hills and trees at the side of the road while the other person finds the way, makes the right turns, looks out for potholes, avoids hitting squirrels or moose. At some point though, you want to take the wheel yourself, or at least offer to drive a couple miles.

Mary exhaled through the side of her mouth and the smoke drifted toward the window with the mobile inside. Her solution was simple and had taken about a minute to come up with. She picked up a scrap piece of the walnut and placed it vertically on the far left side of the deck. It ran from below the platform to the ground, and it would hide the three-quarter inch gap I'd created on the horizontal piece. Just one extra piece of trim, and the deck actually looked better for it. I should've been able to figure it out myself.

"So much of carpentry is figuring out how to deal with mistakes," Mary said. She'd said it before and she'd say it many times again. Her problem-solving ability, the way she could locate solutions or alternative approaches or work-arounds impressed me over and over, and struck me as maybe the most valuable quality she possessed. It comes in part from a brain suited to puzzling out problems in the physical world. It comes mostly from experience. "Half the job is knowing what to do when something goes wrong."

A lot was going wrong. The learning curve had leveled off and the initial exhilaration of the new had given way to the slow climb toward competency with its setbacks and frustrations. More than a year and a half in, I could no longer claim unfamiliarity as an excuse. Some mistakes predate the job: time and moisture have slanted a floor; a previous countertop installer wasn't so concerned with the concept of level; an overconfident homeowner has tried his hand at wiring; walls have bowed, plaster has swelled, tiles have cracked. But some mistakes are of your own making. Many, in my case.

Mary's desire for bigger jobs had been answered a couple times over. We got a big kitchen renovation job on a third-floor condo in Jamaica Plain for folks on a short budget. The neighborhood, south of Boston, inspires deep loyalty in its residents. At the Arboretum, the trees are labeled, and it seems like everyone owns a dog. E. E. Cummings, Anne Sexton, and Eugene O'Neill are buried in a nearby cemetery. One store has over seven thousand hats to choose from, and City Feed & Supply has a general-store vibe, with fancy cheese from Vermont, good sandwiches, and a community-shared commitment to sustainability. It's a sign of the times that the Lucy Parsons Center, a radical bookstore and community space that welcomes all lefty tendencies, moved from Cambridge to Jamaica Plain some years ago.

The condo was bright and airy, with dark and detailed woodwork, large windows, and lots of family photographs on shelves and walls. Nieces and nephews of the couple who owned the place, images from their own childhoods, a woman on a horse looking earnest and focused; another with three kids in snow pants on a sled. The back looked out over a low hill with a beautiful garden shared by the three houses on this tiny dead-end street. The new refrigerator had to be craned into the third floor over the deck at the front—there was no way to get it up the twisting flight of stairs. This was exciting, watching the giant box lifted from the ground, dangling three stories above the sidewalk.

The owners had bought IKEA cabinets, and the whole renovation was being done for about $25k. The job moved along in an orderly way: walls, floors, counters (lovely soapstone, black with streaks of green), and the assembly and installation of those IKEA cabs. It was ten in the morning, we'd been at it for a couple hours, slotting pegs into holes and jamming pieces of smooth laminate together to make cabinet boxes. It was neither boring nor exciting, just something that needed to get done. I moved on to work on a corner cab with a lazy Susan in it. I pieced it together using the wordless directions (sometimes a picture is *not* worth a thousand words; sometimes the right ten words would really, really help). I'd slotted dowels in the right places, secured the walls, top, and bottom, and was drilling to get the Susan to stay where it was supposed to.

But the screw wouldn't spin itself into the material. It wasn't wood, but some sort of plastic-laminate something, white and smooth and as resistant to my drilling as steel to a termite. One after another, the screws skidded off in ricochet, bouncing off the counter, onto the stove, down to the new tile floor with large cream-colored twelve-by-twelve tiles. The plick-plink of the metal screw bouncing down to the stovetop annoyed me at first, then infuriated me. Again and again I pressed the drill into the screw, pushed into the impenetrable composite material, and swore. I swore ferociously. Mary, nearby, looked up from attaching doors to lower cabs.

"Try drilling out for it first."

I heard her but I didn't *hear* her—the advice didn't register, didn't make sense. After all, I was *trying* to drill.

My face flushed. Another screw went bouncing down. I muttered to myself. My shins were sweating. Twenty-five minutes had passed. Not long in the course of the day, but too long to be wrestling with one stupid screw for one lazy Susan. The edges of the shiny screw head, round and flat, bore into the thumb and forefinger of my left hand as I placed it, again, to the drill-bit tip, as though I'd been gripping it there for days, leaving marks in my finger flesh like a too-tight sock leaves rings around your calf. The little screw shined. Light blinked off the metal, reflected the automatic flashlight function of the drill. It was a malevolent sparkle. These tiny enemies didn't deserve such shine. My upper arm pressed against the inside of the cabinet as I wedged myself into position again. My skin adhered to the surface, so plastic-slick, so hard and artificial, with sweat and anger. Each time the screw twinkled off to the floor meant I had to reposition and peel myself off the cabinet surface, which made a shameful suction sound, another indignity. The rev of the drill, the screamy buzz—dentist office, bones and gums—echoed against the walls of the small chamber of the cabinet, where my whole head was enclosed. The thunk of the drill bit hitting the cabinet after the screw blasted off punctuated each effort like an insult. The smell was warehouse, dust and plastic, like Saran wrap, sanitized and dead, and the faint whisper of heated metal inside the warming drill. My breathing grew unsteady, quick shallow pants followed by deliberate slow inhales through the nose, eyes closed, the thudding in my chest a reminder that I was here right now and did not want to be. "The perversity of inanimate objects." It was a phrase my father often said, quoting his grandmother, to describe lids that wouldn't screw back on when you were in a hurry, screw heads too soft for screwdrivers, moments when the dumbness of things outdoes your ability to calmly deal with them. It surfaced on my mind here, the whole situation taking on the air of the perverse.

I put the drill down and reviewed the directions again, with the hope that some new clue would be revealed to me.

"Be smarter than the tools," Mary said from behind a low bank of cabs on the floor. It was another of her refrains. She used it when the tools or the process or the materials were out-smarting us, weren't cooperating the way they should, or we were moving too fast, not considering the best or most efficient way to get a thing done. A reminder that we have brains and the ability to reason and the screw is just a screw. A reminder to take a second and think it out. It was usually a helpful thing to hear. It wasn't now.

I was drilling where the directions told me to drill. I was sure of it. The screw belongs here, I thought. This is what I'm supposed to be doing. What's wrong with the Susan? What's wrong with me? Why doesn't this work? Fuck this screw. Fuck this drill. Fuck this Susan. Fuck IKEA. Fuck me.

Swearing ceased. I was accessing some pre-articulate part of my brain. Huffs, grunts.

"Breathe," Mary said.

I shot her a dark look through the back of my skull. Breathe? I hated Mary and her advice. I hated the cabinet. I hated the tools. I hated the decision to quit my job where I clicked and tapped and got coffee with people I liked. I hated that I wasn't smarter than the tools. That I hadn't learned anything in a year and a half. That my life now involved peeling my sweaty arm flesh off a surface with a vaguely sexual sound. And I hated cheap Scandinavian design.

I repositioned myself again, pressed the drill with all of my body. The screw flipped away and spun to the back corner of the cab, spinning and spinning like an ice skater. I put my head in the cabinet as though it was an oven because I did not want Mary to see that tears had welled up in my eyes.

She passed me a sharp thin drill bit. "Drill out for it," she said quietly. "Make a pilot hole."

And I recognized then what she'd said before. Don't try to get the screw to go straight into the material, make a hole for it first, then drill the screw into *that*. I took the Phillips head bit off the drill and replaced it with the one Mary handed me. I bored a small starter hole into the Susan. I switched the bit back to the Phillips head and placed the screw onto the tip and pressed the screw into the hole and squeezed the trigger again. The screw plunged down and in and the lazy Susan was secured.

I went outside and looked at the lilies in the garden.

Anger had cleaned me out. The hangover from it, the aftermath of all that frustration and embarrassment, left me feeling unfamiliar with myself. I hated Mary a few minutes ago. I regretted every decision I'd made a few minutes ago. What a potent intoxicant, anger. This wasn't the truth, was it? The hangover came with the desire to be alone to reconcile what the anger provoked and what the truth was outside of anger.

In *Wanderlust*, Rebecca Solnit quotes Lucy Lippard writing of an Eskimo custom in which an angry person walks his or her anger off: "The point at which the anger is conquered is marked with a stick, bearing witness to the strength or length of the rage." I have wondered how far anger might take me.

What's to be found at the end of that walk when you plunge your stick into the earth? Anger dissolved in the steps behind you, you return to who you know you are, and turn around to face a new view. So many times with Mary, I would've brushed my hands on my pants and walked away from whatever mistake had been made, given up, resigned to failure with no hope or interest in trying to rectify what had gone so wrong. Board's too short? Might as well call it quits on the whole project. Sub-floor's rotted where the dishwasher leaked for years? Let's get out of here. Mary showed me, over and over again, how a little time and effort, a little care and thought, can correct almost every ill. It's a lesson that translates to love, of course. How many times, after a lapse in judgment, a bad fight, a stretch of boredom, a miscommunication that seemed to signal a total lack of knowing the other person, how many times had I brushed my hands on my pants, checked out, and walked away. It just wasn't working. It just wasn't right. I hadn't learned yet to give it—love—the time and effort it demanded. I hadn't met the person worth the effort. Patience, a little finesse, the ability to stay with something that periodically bored or frustrated you, that periodically drove you to the edge of madness, these were skills necessary too for sharing a life with someone. I do not think it a coincidence that the deepest, strongest love I've experienced began after I started this work. Swearing, yell-

ing, moments desperate with frustration and anger—a break
to walk with a stick, to look at the lilies, and I return, again, to
get at the truth, to try to make it better. Oh I have been slow to
learn. That morning in Jamaica Plain, I could walk no farther
than the garden, and I had no stick to witness the strength of
my rage. But what I saw when I turned again to face the house,
from that view, what I knew, was that I would go back into that
kitchen and try again.

"Try again. Fail again. Fail better," as Samuel Beckett put it.
It's easier to say *screw it*. It's always easier to walk away. Screw
up, and again, try to get it right, and again. Screw up better.

A ll the screwing up got me wondering about the
tool and the word. I knew the hammer was as
old as tools get, and figured the screwdriver followed not long
after. Not so. Not so at all. The famous Phillips head screw
wasn't patented until the 1930s by an Oregonian named Henry
Phillips. Something so simple and ubiquitous—I thought it had
been around for centuries. It's easy to picture peasants slot-
ting big screwdriver-type tools into that crosshatch shape in a
Brueghel painting. The crisscross image is so familiar—Swiss
Army, Red Cross, Band-Aid, Jesus. But medieval peasants
weren't screwing with ye olde Phillips head.

Screwdrivers have been twisted for much less time than
hammers have been swung. Witold Rybczynski, author, urban
planner, and builder of his own house by hand, wrote a whole
book on the humble screwdriver, a tool he named as the most
important of the millennium. In *One Good Turn*, he reveals evi-

dence of screwdriver use from back in the 1580s. But before the industrial revolution it was difficult to produce the metal thread that wraps around the shaft of the screw like a helix. When tools to make the screw were improved—the turret lathe in the 1840s and the screw machines that resulted in the 1870s—screws and their drivers became more widespread.

The Phillips head came about with the rise of power tools. Since the driver tip centered itself in the crosshead slot, you didn't need hands or eyes to align it. In other words, it was good for assembly lines, which is where the Phillips head took off, fastening Cadillacs together on the floor of a factory in Detroit. The stripping of a screw underneath the tip of a screwdriver comes with a thudding, halting sound of error, like you're driving along the highway, smooth and fast, and cement gives way to stones and all your tires go flat.

The word *screw* first appeared as a verb in 1605, coming from the mouth of Lady Macbeth. She urges her husband to summon the toughness to kill King Duncan. "Screw your courage to the sticking place, / And we'll not fail," she says by way of pep talk. The gist is clear, but the precise meaning is not. The *Oxford English Dictionary* believes it to be a reference to tuning pegs on a musical instrument—turn the peg to the sticking place where the right note is found.

The word itself descends from fifteenth-century France. The word *escroue* meant nut, cylindrical socket, screw hole. *Escroue* might be an offspring of the Latin *scrofa*, which means sow, particularly a female pig in the breeding stage. That those two words are linked, *scrofa* and *escroue*, comes down to a wildness of nature: the shape of a swine's penis resembles a corkscrew, twirling at the tip. Photographs show boar cocks thin and spiraling, and the female's cervix is ribboned the same spiral way to lock the penis into place when mating. When pigs are inseminated artificially with pig sperm, the tool often used is called a spirette, a long thin rod that spirals at the end to mimic the swine penis shape, and inseminators twist it in counterclockwise. In Iceland, the word for screw is *skrúfa*, and also means to fuck.

Pig private parts feature in other etymologies. Cowrie shells, egg-shaped and white smooth, have a rounded back and an opening slit on the front. They were called *porcelaine* in French, *porcellana* in Italian, the diminutive for the young and fertile sow. The shape of the shell is said to have called to mind pig vaginas, and so it got its name in Italy and France. Porcelain—chinaware—resembled the shells in its smoothness, and so got its name, too.

Cops are called pigs, and in mid-1850s England, screw was a slang term for a prison guard. Two theories exist as to why. One: screw was another word for key, and any image of a prison guard involves a jangling ring of keys hanging at the hip off a belt, or being slapped ominously against the hand like a tambourine, the clanging rattle a reminder of who has the power to lock and unlock shackles and cell doors. Two: jails in mid-nineteenth-century England were places of punishment, one of which involved prisoners cranking on a handle that turned nothing but a counter. Turning the handle ten thousand times in eight hours was a common penalty, and as the cranking continued, the prison guard would tighten a screw to increase the resistance, making it harder for the prisoner to rotate the apparatus. It could also be reference to the guards torturing inmates with thumbscrews, known nursery-rhymishly as pilliwinks, which were simple vises that were clamped and tightened to crush someone's thumb, finger, or toe.

To get screwed—cheated, hornswoggled, taken advantage of—might be an offshoot of the prison-guard slang, developing in the criminal underworld and making its way to the street. In French today, *écrouer* means to imprison. *Levée d'écrou* is the release of a prisoner, literally translated, the lifting of the screw.

The addition of "up," William Safire explained in a 1990 *New York Times* column, came out of a World War II lingo heavy with euphemism for botch jobs and errors: gum up, foul up, mess up, and screw up. The latter appeared for the first time in *Yank*, a weekly magazine put out by the United States Army from 1942 to 1945. Safire made the claim that Holden Caulfield had a hand in popularizing the phrase in J. D. Salinger's *The Catcher in the Rye*. "You know what the trouble with me is? I can never get really sexy—I mean *really* sexy—with a girl I don't like a lot. I mean I have to *like* her a lot. If I don't, I sort of lose my goddamn desire for her and all. Boy, it really screws up my sex life something awful. My sex life stinks."

M y sex life didn't stink, but I was finding that the carpentry work was altering it. The work I was doing didn't make me less of a woman, but it felt like that, in a profound and surprising way.

It's odd to admit, and I feel small for doing so. I started noticing something shifting as I got dressed in the morning. I slid my legs into paint-stained jeans, crusty patches around the pockets where I'd wiped glue from my fingers. I tangled into a sports bra, pressed breasts, pulled on a tank top, a T-shirt, a long-sleeve shirt, weather depending. My sneakers were worn, with gray-white dried cement caked on the toes, paint blotches, more glue. I tied my long hair back into a bun, made sure I had my earplugs, and headed out into the day.

I looked grubby and thick. I looked like someone doing manual labor. I felt like a boy.

I am not a small woman. I am sturdy and curved. I feel lucky to have avoided the body image demons that skew reflections and make some women loathe their flesh. I like that I am strong. I like the muscles in my legs, quads and calves, and that my legs can run for miles. I flex my biceps in the mirror and feel proud of the bulge and the strength it suggests. I love having tits. I like the combination of firm—legs, shoulders, back—and soft—breasts and belly. I like the blurring, the strength and softness. As Virginia Woolf wrote, "It is fatal to be a man or woman pure and simple; one must be woman-manly or man-womanly." This feels deeply true to me, an abiding desire to have both sexes mingled in one body and one mind, a mental fertility.

But dressed in work clothes, breasts bound in a sports bra, spending the days doing work with my body and hands, holding tools, tape measures, hammers, using saws to slice wood, using nail guns and drills, it shifted my sense of myself, my sexual self, my self as a woman. I hate admitting this. I hate that dirty jeans and using a drill were enough to disrupt my sense of my own self as a woman. I felt desexualized.

In this way, I moved through the world without projecting that spark of possibility. In being dressed in work clothes that obscured curves, absent was that energy I gave off and got returned from men around me, absent was the sense within myself of sexual desire. I noticed not noticing. And I noticed not being noticed. The newsroom where I'd worked had sizzled with flirtation, crushes galore. I'd worn jeans and tight turtlenecks and had not worn makeup.

I started wearing mascara and eyeliner when I was thirty, some months after I'd signed on with Mary. I didn't realize it then, but it was balancing. There were some days I longed to come home, lean in to kiss my boyfriend smelling of perfume instead of sawdust and sweat. In my non-work hours, in an effort to even out the feelings I was having inside myself from the clothes and the work, I'd get home, shower the sawdust off my skin and hair, put on tight jeans, a lace bra, a low-cut shirt, and brush mascara on my lashes and rim the lids with liner. I'd always loved watching women put on makeup in public-bathroom mirrors, and found myself enjoying the practice of it, learning how to do this at thirty instead of thirteen.

With the sexual switch turned off in work clothes, I tried to turn it on in non-work hours, upping the feminine in ways I hadn't before. I found that femininity and my sexuality were tied together in ways I did not expect. I was jarred that outward signifiers altered something inner. I found myself, when the plumbers and electricians were around, mentioning boyfriends, past and present, making them know, deliberately, that I was a woman who was into men. It felt forced, and I wonder if they felt that, too.

In the off hours, I imagined the plumbers on top of me. I thought about their big arms, their rough thick fingertips. I hadn't tried to flirt. During work, it was as though I was nine years old, a return to a time before I was animated by sex. But after work, there they were in my thoughts. I imagined the weight of them, how the strength of their hands and arms would translate to flesh. I'd see the older one we worked with, the bald-headed one, tall and strong, the length of him outstretched on a kitchen floor, reaching up under a sink, gripping a wrench—during the day, it was a fact of the job, his specific task, sexual as a pack of shims. Later, at home, away from work, the thoughts turned close and humid. At the job the next day, it was back to the chasteness of childhood, as though the imaginings from the evening before hadn't existed.

In the day-to-day work with Mary, I didn't think about the numbers either—the percentage of women doing this work didn't occur to me. I didn't chop boards and maneuver sheets of plywood and shoot the framing gun contemplating how few women do this work, fewer still heterosexual women. Here we were, Mary and I, both of us strong, one of us capable, and together we were able to do what needed to be done.

The fact is, carpentry is men's work. Which is to say, carpentry is work that is statistically done by men. The U.S. Census Bureau, in its 2011 survey, reports that "construction and extraction occupations" are made up of 97.6 percent men and 2.4 percent women. It is the most gender disbalanced of the occupations they list, more than engineering and architecture, more than farming, fishing, and forestry work, more than firefighting.

Some estimate an even sharper discrepancy. In an article in *Monthly Labor Review* called "Gender Differences in Occupational Employment," Barbara H. Wootton notes that "the most pronounced differences in occupational employment by gender occurred in precision production, craft, and repair occupations—in 1995, for example, only 1 percent each of auto mechanics and carpenters were women."

It's not a statistic that seems to be changing much over the years either. The Washington, D.C.–based think-tank Institute for Women's Policy Research (IWPR) reports the shifts of women in the workforce in a paper titled "Separate and Not Equal? Gender Segregation in the Labor Market and the Gender Wage Gap." They tracked the rise of women in the workforce from the early 1970s through 2009. Back in 1972, only 1.9 percent of dentists were women; in 2009, women made up 30.5 percent of the profession. During those same years, the percentage of women mail carriers rose from 6.7 percent to 35 percent. In the trades, though, the numbers haven't changed much. Female carpenters made up 0.5 percent of the workforce in 1972 and, as of 2009, made up only 1.6 percent. Carpentry also happens to be among the whitest professions. In a November 2013 article, *The Atlantic* reports that carpenters are 90.9 percent white and notes that trade unions have had a "complicated, and often ugly, history with race that's helped shut blacks and Hispanics out of these highly coveted lines of work."

Susan Eisenberg's book *We'll Call You If We Need You: Experiences of Women Working Construction* documents the experiences women had on jobsites in the late 1970s and 80s. Part oral history, part documentary, the book details the harassment and disrespect that women experienced working in the trades. In it, a woman named MaryAnn Cloherty describes having finished a nine-month pre-apprenticeship program and approaching one of the carpenters' locals. "A hiring agent said to me, point blank, 'We had the colored forced down our throats in the 60s and we'll be damned if we have the chicks forced down our throats in the 70s.'"

Eisenberg, who herself is a master electrician, also details the pride, passion, and satisfaction these women experienced, as well as some of their generous, patient, and welcoming male colleagues and mentors.

The IWPR cites the "hostile environment" in many male-dominated trades as a reason why so few women have access to these jobs. "There is considerable research suggesting that occupational choice is often constrained, by socialization, lack of information, or more direct barriers to entry to training or work in occupations where one sex is a small minority of the workforce." If no other women you've known, or even know of, do a certain job, it won't necessarily feel like an option for you as you take steps to plot out what you might like to pursue in your own life. And I think just as there are certain jobs (nurse, dental hygienist, secretary) that some men might feel would call into question their masculinity, the same goes for women. There are certain jobs that raise questions about femininity. Though I didn't often reflect on the scarcity of women doing this work in a general sociological sense, I did find the work challenging my own ideas and sense of femininity and sexuality.

We got noticed. Selecting boards at the lumberyard, or loading up a cart of drywall, or hoisting sacks of cement at Home Depot, surrounded by big construction guys in their coveralls and clomping work boots, the looks we got weren't just curious. "Got a little project going?" asked one orangeaproned Home Depot checkout clerk, as though we were four-year-olds gluing popsicle sticks to construction paper.

"Renovating a kitchen," Mary said matter-of-factly, pulling her credit card out of her wallet. She showed no trace of defensiveness or aggression. I hoped the glare I gave got those things across.

Now and then, a lumberyard employee would address her as sir. "Help you find something, sir?" I've been tempted to shout "You mean *ma'am?*", defensive on her behalf. I haven't said it—Mary can fight her own fights. But I'm not sure she considers it a fight at all. "All set," she'll say, unfazed. In third grade, I got a short haircut. In school the next day, I held a door for a teacher, and she said "Thank you, sir." I was silenced, reeling. *Sir?* She clicked past in her high heels, and I felt shaken, chaos in my head. What am I? Am I not what I thought I was? The teacher's comment stripped me of my understanding of myself. I can still feel my small self standing in that doorway, can still conjure the confusion and the fear. Every time Mary gets mistaken for a sir, I am holding the door in third grade, tipped upside down by being mistaken. For Mary, it seems not to matter. She's less interested in belonging in a feminine category. And I am more attached to my femininity than I'd known.

Another morning at the lumberyard, I saw two young guys in thick Carhartt jackets as we loaded planks of one-by-four Ipe onto a cart for a deck job. One gestured at us with his chin and said something to his buddy in a whisper, and they laughed like middle-school girls. I don't care to imagine what was said. My initial red-cheeked impulse was to grab a plank of iron-wood and whack them across the shins. Instead, as we passed by them, I said, "Hey, boys," in a purring tone and raised a provocative eyebrow.

When we were out in the world, I was on alert for skeptical glances and condescending remarks. In more generous moods, I'd remind myself that not all these big contractors with the pick-ups and muscles were assholes, that it *was* unusual to see two women loading lumber. The guys Mary hired knew her, they'd worked with her for years, they were used to being on a job with a woman. Not many men are. So, yes, you want to stare: go ahead. I figured our presence there, loading the cart, pulling sheets of plywood off the pile, stacking up pressure-treated four-by-fours, lifting sacks of cement, might open one or two of them, even for a moment, to the possibility that women do this work too.

There were two moments I can think of when we prevailed upon bigger, stronger men to provide some extra muscle. One was getting a hand to move a sliding glass door up to a third floor. Mary's old boss was working down the block at the time, so he and one of his guys played big strong men for us. Had there been two women around, with strong muscles and good spatial sense, that would've worked, too.

The other time, we were rightly thwarted. Building a big new deck for a house on a hill in Jamaica Plain, we spent a day digging the footer holes, each one two feet wide and four feet deep to reach below the Massachusetts frost line. We dug and dug, took turns with the post-hole digger. We sweat and dug, and the sun, on this south-facing house, creamed us all day long.

It smelled like onions when we dug. We excavated bulbs and with our shovels charged through roots and small clusters of chives. Lemon drifted from a small patch of lemon balm that we had to stomp over, right in our path to and from the van in the driveway. Our footfalls broke the leaves and oils leaked and the smell—lemonade, citrus fresh, a soapy soothing—spread. Lemon balm is said to improve mood and mental performance. Whether that is true, I don't know, but it was a welcome alternative to the smell of dirt and sweat and suntan lotion, the closest odors always on those hot days. I learned those weeks that drinking is one cure for confusion and fear, a temporary relief from what feels impossible to face. Digging holes in the dirt with the smell of onions is another. I was grateful to go to work those days and lose myself in the labor.

It was a Friday morning, and the heat was something to move through. We were about two and a half feet deep into the earth when we hit a rock. We'd hit rocks before. We'd keep digging around it until we got it out. We dug around this guy, but it was large and crammed in and we couldn't find its edges. We used a crowbar, a shovel, a winch, a sledgehammer. We used all of our muscles. We tried the canvas straps that Mary used to tie her canoe to the roof of the van. We couldn't budge the rock. We toiled and we cursed.

"It's just one big rock," Mary said again and again, to console us, maybe, to assure us that we weren't the weaklings here, that we'd encountered something in the earth bigger than we could deal with.

As we dug and shook our heads and tried to bash the rock to pieces, three men in hardhats and work boots excavated the sidewalk and the street nearby to repair a pipeline below the cement about a hundred feet from where we were working. They used an excavator to rip up a strip of the road, clawing through the cement, pushing debris into the loader. The trucks roared and the noise made the air feel hotter. Mary made a comment to the guy driving the big machine, something about lending a hand with our rock. He shrugged in a way that said no-can-do.

So Mary decided the only way to deal with the situation was to rent a jackhammer for an hour and break the rock that way. I did not like this idea at all. I pictured big men on work crews, their arm flesh jiggling, their teeth rattling in their jaws, riding the jackhammer like a bucking donkey. I did not want to try this.

"First time for everything," Mary said out the window as she drove off in the van.

I started digging another footer when one of the construction guys came over to me. He had thick freckled arms and the hair on his big shoulders was starting to whiten. His jaw was scruffed with white-blonde stubble, and he wore his orange mesh work vest with nothing underneath. I could smell his sweat and I liked it. He asked about the progress and told me I should drink water. I laughed and told him he should drink water, too. I told him about the rock and pointed into the hole.

"That's a big one," he said. We stood in the sun and looked into the hole, both of us dirty and sweaty, hands on our hips, and I made a decision. Sometimes big strong men want to feel like big strong men and I said something that I didn't want to say but I said it anyway.

"I guess we're just not strong enough."

And he looked at me and said, "We'll help you get that rock out."

He walked away and pointed up at me to the man driving the excavator machine. The excavator machine rolled up toward me, I jumped out of the way, and the shovel bucket went clawing into the earth and scooped up our giant rock just like that, a boulder, two hundred pounds or more. I thanked the guys and they seemed happy, too.

I called Mary to tell her we didn't need the jackhammer. When she got back, I told her what happened and she laughed and thanked me for showing a little leg.

The next part of the project underlined our toughness. Once the holes were dug, we sank the footer tubes, which look like enlarged versions of the cardboard tubes we roll posters into, into the holes, and filled in the sides with lose dirt. Then came the cement. Each footer tube needed to be filled, as did the pad at the base of the stair.

We ripped open the bags, held our breaths, shook the rocky dust and sand into a big plastic tray, and used the hose to wet it. Each of us with a shovel, on either side of the tray, we mixed the cement, bag by bag. We pushed it, churned it with the shovel edges, heaved it back and forth, wetting it evenly, not too much, not too little. I wore a mask; Mary didn't. Neither of us thought we'd get through it in one day. Work sometimes had the feel of summer camp; it felt simpler, and more like play. Whether the obscuring of sexuality contributed to this summer-camp feeling or was caused by it, I don't know. But at times it felt like a traveling back to a time before I had breasts.

We'd finished the four footers, and the sun hadn't yet even reached the trees by the driveway that signaled the three p.m. shade. We had some hours left in the day.

"What do you think?" Mary said.

I took my mask off, used my shirt to wipe sweat and grime off my face, spit on the sidewalk and said, "Let's keep going."

"Hell yes," Mary said.

So we pressed our shovels back and forth, and light made rainbows off the spray of the hose. The gravelly sound of rocks and sand being pushed around the tray sounded like rocks getting rolled under waves at the beach, and shifted to a wetter slop as Mary added more water. "How many bags do you think it'll take to fill the pad?" she asked.

"Six?"

"I'd say double that."

We mixed and poured, mixed and poured. A blister blossomed inside my work gloves, ripped, and leaked a sticky warmth into my palm. The stair base started to fill. When it finally reached the top of the wooden box, some slicking down the sides in slim rivers of gray, Mary and I high-fived. All told that day we loaded, unloaded, and mixed two thousand five hundred sixty pounds of cement. One and a quarter tons.

"Unreal!" I said, exhausted.

"Not bad for two chicks."

"Not bad for two anything."

A n encounter on a kitchen job did, finally, inject sex into the work. The job was in Framingham, a town twenty-three miles southwest of Boston, farther afield than we typically traveled, but work was work despite the forty-minute ride to and from. The house sat nondescript at the end of a suburban cul-de-sac. Each house looked the same; paint color, shutter style, and choice of front-yard shrubbage were the only distinguishing features. We went through the usual process: tiled the floor, installed the cabs, and wrestled with crown molding. Much of the work was done when a handsome granite countertop guy named Pete arrived in his truck to take measurements and make templates for the piece of stone he'd be cutting for the counter.

He had dark curly hair and an easy smile. When he leaned over the counter to make a measure, his shirt settled into the runnel his muscles made along his spine. He had arms like a discus thrower in a Greek statue. I had broken my wrist earlier that season, doored by a young woman in a BMW as I rode my bike home from work one afternoon. This was my first job back after the cast had come off, and I wore a black brace on my wrist as I worked. Pete asked about it. He talked about ripping his Achilles not long ago, and bragged.

"The doctors told me it'd take eight months to a year to heal," he said as he pulled the tape across the top of the cabinets. "You know how long it took? Three months. I was back in three months. If you know your body, you will heal. You will get through. You have to know your body."

I liked that we were talking about bodies, about knowing them, something immediately physical and intimate. I repeated something my wrist doctor had told me about pain, how it can be a good thing sometimes. He let his tape whip back into its holster as though I'd said something he'd been waiting to hear. And he stood with his back to me, that strong back, and said, "It shows you're alive." He turned to face me, looked me in the eye and said, "We need to know that sometimes."

And then he winked before turning again to measure where the sink would go, and it was too much and too smooth, but I smiled despite myself and my stomach dropped into my hips in that warm pulse, and I looked forward to him coming back with the granite slabs. How far these tiny moments of heat can go, a flash of the eyes, an elevation of the atmosphere, these brief pulses of shared intimacy, of energy passed back and forth. Nothing more than a conversation, less than a minute of talk. I wasn't going to fuck this man on the smooth cold granite stone he'd bring, but I thought about it.

The next week, granite cut and ready to be slid into place, he came back to the kitchen. I flashed eyes and a smile at him like I'd done a thousand times at bars, to boys and men, friends and strangers. And he aimed it right back and said good to see you. It was nothing, the most basic interaction with another human, but the energy was there, that flash back and forth. And in that moment, I came to know the flash was just as strong, that power and its return. All it required was accessing the energy I had when I wasn't in ragged sneakers with a chisel in my hand. He might've flashed eyes at anyone, likely did, this curly-haired countertop man, but it did the job of showing me that even in work mode, I could aim the energy and have it returned. And some sense of fullness returned, an embodiment—physically and mentally—of Woolf's conception of woman-manly.

There is screwing, and there is screwing up. I did so much of it. Mistake after mistake. After the kitchen job with the countertop guy, we moved on to a job in Lexington, where the crosswalks are strictly enforced and tour guides in minutemen costumes lead history buffs around important historical sites. We were there to redo the first floor of an old carriage house—new floors, walls, kitchen, bathroom, some new windows, lots of doors and trim. It was a big job. The place edged up against an old graveyard filled with tiny headstones from the 1700s. One section of wobbly graves described an almost perfect circle, which created an additional layer of haunt and gloom. Every hour, one of the minutemen came around and showed his tour group a grave right by a big window where we were slamming in new floors. He wore full colonial-militiaman regalia. The thought of these guys climbing into their cars at the end of the day, placing their triangular hats on the seat beside them, made me feel sad. There's something lonely about existing in two times. One damp afternoon, the tour guide was holding a can of Mountain Dew. What struck me wasn't the anachronistic clash. It was more a sense, out of nowhere, that you shouldn't drink soda in graveyards.

Mary had me trim the inside of a bedroom closet, an odd trapezoidal shape on an uneven floor against a bowed wall. I puzzled through how to get the correct angle of cut for a piece of baseboard that would line the section that ran from the side of the closet door to the corner on the right. This is a piece you would see if you stood inside the closet facing out into the bedroom. This was not a walk-in closet; it was a regular closet that you'd reach in to grab your button-down or your corduroy dress off the wire hanger. Unless you were hiding in there, there would be no reason ever to see this piece of wood.

I was trying to make the piece flush against the floor, flush against the wall, and flush against the other piece of trim it bumped into in the corner. These are always the goals with trim. Some days are easier than others.

I walked from the closet to the garage where we'd set up the saws. Over and over I traveled this path as I cut and re-cut the pieces of trim—across the new hardwood floors, through the kitchen, past the small bathroom under the stairs, and out into the garage. Frustration flared. Who cares how this piece fits? No one will ever see it. Such a waste of time. My motivation faltered—oh, it's good enough like this, isn't it, with a gap between the wood and the floor? Leave the wide seam there in the corner, just slop it up with some extra caulk.

I made half-blade cuts, shaving off pieces of the wood, one angle degree at a time. I took a degree off the lower edge of the right side where it would hit the other piece of trim in the corner. It fit snug and tight. The floor bowed so that the right and left sides of the piece knocked back and forth like a seesaw. I lay on my side, legs out the closet door, and ran my flat fat pencil across the length of the board following the rise and sink of the floor. The line on the wood that resulted showed me where and how much the floor swelled and where I needed to remove the wood. Shave and shave, and finally it fit flat.

"It takes years to get good at this," Mary said as I walked past her toward the saws with my piece of trim, shaking my head.

When two pieces met in a perfect seam, when the pieces followed the bowed swell of the floor just right, gapless, steady, when it pressed in to fit tight and right, I exulted. The simple joy of it!

Whether Mary was trying to teach me a lesson that day, I don't know. Part of me thinks she knew it'd be tricky for its bows and angles, and trickier still for its out-of-the-way-ness and won't-be-seen-ness, a test of technical skill as well as something mental. Or maybe it was just another thing that needed doing, and she had more important projects. She heard me swearing. She saw me walking back and forth. She stayed quiet, let me puzzle my own way through. I got pieces wrong. I'd make one extra slice and take a little too much off and render the piece useless. Our board stretcher was always at the shop.

But if she had come to the closet, seen the fat gaps and teetering pieces and said, "Yes, to hell with it, it's a closet, who cares," I don't know that I'd have felt much relief. Maybe in the short term—*thank god I'm out of that closet*—but it'd have been cheating. If you are able to maintain focus and attention for a piece that will not matter, that will rarely, if ever, be seen, if you are able to get that right, the rest of the work—the stuff that does matter, that will be seen—will be elevated.

During a trip to or from the saws in the garage, something shifted. Impatience changed to purpose, to mission mode. I went from being faced with a tangle and ready to throw it across the room to being able to see both ends and following the strands out of its snarl, slow, patient, precise, until one smooth string could be held and stretched, glowing between two hands. This will be right. I will make this right.

Were the eventual owner to look inside the closet, I'd bet he or she would not notice if it were done well. Done badly, though, with gaps and slop, it would come into focus and raise questions about the quality of the rest of the work around the house. An uncritical eye learns to see what can be wrong, what's done sloppily, lazily, without effort.

Trim on the inside of a closet *does* matter. It might someday belong to a messy teenager who heaps sweaty shirts from soccer practice, old socks, damp and sandy beach towels, ragged school notebooks, so that piles hide the trim. So what? The pieces will disappear. That's what's supposed to happen. It wouldn't have kept me up at night knowing I left a thick gap that required a smeary spread of caulk to cover up. But the satisfaction, the quiet sense that I got it right, that it mattered, made it worth it. And whether Mary had intended it or not, it made me glad to have done it.

"Finished," I said to Mary, who was framing windows in the main room.

I didn't notice when she put her tools down and walked into the bedroom where I'd been. But when she came out a few moments later, she gave me a thumbs-up and nodded her head. She walked back to her tools and said, "You want to start trimming the dining room?"

I made my way with the tape to a corner by a window that overlooked the graveyard.

There's no backspace key in carpentry, no control+Z. You cannot refresh a miscut piece of wood. I took for granted the undo-ability of labor in my old job. A couple quick clicks and anything could be fixed. Correcting for the errors in carpentry involved a new set of brain skills, ones that do not come naturally to me, and ones I was so grateful to be gaining.

I botched a simple chiseling job working on the same carriage house. I needed to carve out the space in a door where the hinge would go. I'd traced the outline of the hardware on the side of the door and began digging in with the chisel, aiming for an even one-eighth-inch depth. The wood curled off like ribbon underneath the press of the chisel edge, paper thin, and fell to the floor without a sound. I liked those curls so much, I kept going. Mary came in as I straddled the door between my legs, pressing off another shaving of wood.

She placed the hinge against the space I made and shook her head. It sat too deep by an extra eighth.

"This isn't going to work," she said. And she gave me a container of gray-brown wood filler to smooth into the space I'd hashed so I could start again and try to get it right.

But look at those curls, I wanted to say. *Look at how right they are.*

Smearing the gunk into the notch felt like a corruption, like I was defiling the pure wood with something chemical and artificial. The goop dripped and stuck and stank. It did not cooperate. I smeared it this way, back, diagonal across the small space where the hinge was meant to go. Mary came back in and leaned over my shoulder.

"It's not icing," she said. "You're not working with cake."

I said okay, and as she walked away she said, "Sometimes the most important thing is knowing when to stop."

Chapter 4

———

CLAMP

On the necessity of pressure

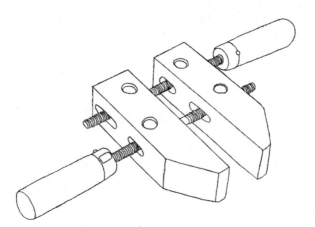

M ary and I slipped into our third year together. Jobs presented themselves—bathrooms, kitchens, decks, bookshelves—and we took the work. The rhythm of the days felt natural now, the rhythm of the jobs, familiar. Fall inched toward winter, and the heap of job debris in Mary's backyard showed the history of our months and months of work. Every time we finished a job, we unloaded bags and bags of trash from her van onto this pile on the side of her yard by the fence. It was now the size of a subway car.

"I've got to get rid of this before it snows," Mary said.

She called the demo guys, the same ones who took down her chimney, to come remove the pile.

On a morning in November, the three men arrived, dad and his two sons, thudding out of their truck. They stood by and assessed the pile: metal pipes, strips of drywall and cement board, a box spring, a pallet, two-by-fours, two-by-tens, wire backing to an old tile wall, wood scraps of varied length and thickness.

"We're looking at five tons here," said the leader, smiling under his thick mustache. Five tons struck me as an impossible weight for three men alone to load in a day. The blonde son, the lean one with the empty eyes, climbed on top of the pile. He stood there, hands on narrow hips, king of this trash mountain. He bent, picked up a group of long boards, thick nails spiking out, and tossed the wood into the back of the dump truck with a metallic clamor. And so it began. Once they started moving, they did not stop. Wood thunked on the bed of the truck. Drywall pieces broke in dusty fragments. Trash bags flew like feather pillows.

The father once again let his sons do most of the hefting and heaving. They shouldered and tossed while the boss discussed his loading system, practiced and refined over years of experience. Flat and wide go first into the deep bed of his truck, long planks next. Stack them neatly, one direction. Metal in another corner; it doesn't go to the dump—there's big money in scrap metal. The odd stuff and strange shapes go after that, with heavy bags on top to keep it all in place on the road. When one of his sons placed a bag of lath, dusty and full of rusty nails, upright, the boss chided him. "Hey, hey. No," he said. And he explained where it should go and why. He said it without anger or impatience. He just wanted it right and wanted his son to know why it was right.

After detailing his loading process, he talked about the merits and drawbacks of area dumps.

"The place over there, you don't want to go there. You know why? Because they don't care. There's going to be nails on the driveway. Do you know what I mean? That place, I'm telling you, it's a mess, and what you do is you risk a flat, you risk a puncture every time you roll in there. You don't want to go there."

Another place nearby takes anything, he told me.

"For them, it's all about the weight. They'll take it all," he said. "Anything. I'm talking about dead bodies."

"Come on," I said.

He looked at me and his face was grave. "You think that doesn't happen? You think that doesn't happen? It happens."

Five tons did not turn out to be an impossible weight for three men to load in a day. In less than an hour, every board and bag had been loaded onto the truck, Mary's backyard emptied, the earth raw underneath.

The boss waved his arms at the load.

"You see all that? Tomorrow it'll be five hundred feet underground in Bangor, Maine."

It felt like I was being let in on a secret I didn't want to know. Trash in a grave, dead bodies at the dump, the remains of our jobs under the ground, decomposing, leaching into the earth, doing damage impossible to see, the same damage it does to our bodies as we work. A familiar feeling of unease returned.

A s the jobs grew familiar, so did the fears that rolled in before I fell asleep—about the dust and spores and toxins we were exposing ourselves to in the work. Eyes closed before sleep, the last things I often saw in the darkness behind my eyelids were dust particles, shifting in the light, dancing in the air in a way that spoke illness and menace. When I coughed I'd think, *Here is the first sign of the tumors growing on my lungs.*

Mary thought of me as a worrywart and teased me gently

about my nerves. That was fair. She told me to take a look at the guys we worked with who never wore masks either. She also indulged my fears, which was a great kindness. When we worked at her basement workshop, we'd sand outside if it wasn't raining. She'd often mix mortar for tiles because she knew how I hated the dust. No doubt she wished that she worked with someone less fearful, because precautions can slow things down. I worried for her, too.

She made the argument, half jokingly, that smoking protected her lungs from worse poisons. She rarely wore a mask. "Nothing's going to get in there," she'd say. "Why do you think I've smoked for all these years?"

Part of me bought this magical thinking.

I envisioned her lungs coated with something black, shiny, cratered, firm to touch. And in microscopic vision, I imagined particles sucked in, tiny pricks of fiberglass, arsenic bits from the pressure-treated wood, formaldehyde from the plywood glue, mites of cement that start to cure when mixed with water. I saw these bits float into her blackened lungs, drift around, and bounce off the black, entry denied.

No protection is gained from the tacky tar in her hand-rolled cigarettes. Despite knowing this, it made me feel like my own pink lungs were even more vulnerable, lacking that black shield.

I had the habit of reading warning labels on all the products and materials. Somewhere in the safety section of almost every label were the words: *Contains chemicals known to the state of California to cause cancer.*

"Good thing we're in Massachusetts," Mary would say.

She talked of doing some insulating on a small project. She said she was going to break down and get a respirator. "I can't do it anymore, with the insulation. It gives this prickly rash around my mouth." She joked about getting me a hazmat suit and I told her I would like that very much.

"See? You don't need to worry," she said. "I get a rash now. Your body has a way of telling you when there's something wrong."

I didn't say anything about the body's ability to keep dark secrets, too.

Once, on a late summer afternoon, a few months before the November morning of the demo men, we sat and watched birds dip and wing in the garden of a client's backyard during the break. Celine, the owner, joined us outside. She remarked how I always wore a mask and Mary never did.

"You're being exposed to so much. You can't even know what you're being exposed to," she said. She made her own yogurt so she could store it in glass containers instead of the store-bought plastic tubs that leached chemicals.

"Something's going to get me," Mary said. She shrugged off the issue with her usual no-nonsense nonchalance. But I sensed a deeper sort of resignation. She was acknowledging that *yes, someday I'll die, I do not know how or when*, but there was something dismissive, too—*perhaps a polar bear will eat me, or a man-eating slug*. She pinched the burning tip off her cigarette and mashed the ash with her foot. "As long as it's not my lungs. That's the one thing I don't want."

What? I wanted to shake her. You don't want it to be your lungs? Be smarter. Smoke if you want to smoke, but wear a goddamn mask! A shyness followed my initial disbelief and frustration—it was an honest, vulnerable thing to say, and it surprised me.

I wondered if the dust clouds ever appeared to her in those moments before sleep. I wondered if she worried about her cough. Wood and mortar dust and fibers and smoke and tar, these were Mary's pursuants, the things that will catch up with her, the things that will get her in the end. She takes them in. Maybe it's not a death wish, but something closer to the sensibility Joseph Conrad captures in "The Secret Sharer." A man throws himself overboard and escapes the ship; his crew think it suicide. "Let them think what they liked, but I didn't mean to drown myself. I meant to swim till I sank—but that's not the same thing."

The demo guys left that morning, and the empty space and the raw earth in Mary's yard marked the unofficial end of the season. From mid-November until the end of the year, work slowed to a halt. People do not want chaos and mess in a time of chaos and mess; hammerbangs do not make the best accompaniment to Thanksgiving feasts or Christmas cheer. Mary and I tied up a few things in her basement, tucked tools away, slotted boxes of screws into bins, swept and neatened. We stood chatting and I held the Jorgensen wood screw clamps, tightening them and loosening them as we talked, pedaling my hands to make the wood tips press together, tighter, tighter, then release. Made of maple and steel, the clamps are tightened by cranking on the handles as though pedaling a bike with your hands. They are powerful, able to eliminate even the smallest space between two pieces of wood when tightened. The power of them, how they erased space, came as another surprise—so simple yet so strong.

"Nothing lined up for the next little bit," Mary said.

I pedaled the clamps, tightening again. "Call me when you need me." I hung the clamps on the pegboard wall, let them dangle with the rest of the tools. The word comes from the ancient German word *klam*, which meant to press or squeeze, and the tight-closed shells of the bivalve, like hard lips locked in silence, got its name from there.

I anticipated the slowing, the early winter hiatus. It was a pause for breath, with a quick odd job here or there, until things started up again in earnest in the new year.

We parted ways that day, wished each other happy Thanksgiving and good luck for handling holiday mania, and said we'd talk before Christmas.

That winter brought huge snows to Boston, more inches every week. The new year arrived and I waited to hear from Mary about the next big gig. But no call came. I left messages on her voicemail: *Hey Mary, just checking in, seeing what's cooking these next couple weeks. Give me a call.*

I got no word back.

The days were short and full of snow. High snow banks narrowed streets, and great battles for parking spots flared across the city. Traffic cones and folding chairs marked saved and shoveled spots.

I read and wrote and took long walks in the snows—there

are worse ways to spend the days. I stayed out late, said sure, I'd love another beer. I had nothing to get up early for, after all, didn't need my body to be rested or my head clear.

I languished, got softer. Muscles strong from carrying saws and swinging a hammer and pressing on a drill weakened, got slack from lack of use. Days disintegrated.

Fallow periods are something to savor. Times of low productivity can be one of life's luxuries. Though there might be no outward proof of action or making—nothing written, nothing built—such time is hardly wasted; puzzles are explored and problems solved in the head. And these quiet times give me a chance to scrape off the emotional muck that accumulates and coats my brain over time. Fields are left fallow, after all, to make the earth fertile in future seasons. Just because we can't see the cornstalks or the swaying wheat doesn't mean that nothing valuable is happening there underneath the surface.

In a poem called "The Summer Day," Mary Oliver writes of strolling through fields and kneeling in the grass, of being "idle and blessed." "Tell me, what else should I have done? / Doesn't everything die at last, and too soon?"

She wrote the poem in 1990, before most of us had cell phones. A couple decades later, in our frenzy of scramble and dash, who has the time to be "idle and blessed"? Who but poets fall down in the grass? Strolling the field may seem lazy, but think again, Oliver says. Who knows what might happen in the stillness? Who can guess what you will come to know eye level with the grasshopper? And she gives us the most crucial reminder: "Doesn't everything die at last, and too soon?"

That's the grass. That's the grasshopper and the fox and the flower. That's you, too, and me.

We can't clutch every moment, but it is good to step back and consider our plans. To kneel in the fields, to laugh with pals at the bar, to look at the swirling grain of the floorboards. This is not original advice. But for those fallow periods to feel both purposeful and luxurious, they need to be bookended by accomplishment, by doing and producing.

But this pause in carpentry was not a fallow period. It did not have that fertile feel. The ability to spend nine days on something that should've taken two, or one day on something that should've taken three-quarters of an hour, provoked a feeling of uselessness. It dredged the question, again and again, *what now?* I did not know when work would come again, if work would come again, and I allowed the fear to keep me from translating this quiet time into something productive or worthwhile.

*T*he more you do, the more you get done. I don't remember when I learned the adage itself, but I know when I learned the truth of it. My dad lost his job in 2001, a few months after I graduated from college. I no longer lived at home, but my young brother relayed what the scene of this sudden unemployment looked like. My father, at the computer, logging hour over hour playing backgammon and solitaire, reading online fishing forums, scrolling idly. A constant clicking from the office, which in the evenings was punctuated with ice cubes plinking against the glass as he took another swallow of Scotch.

There were efforts at first. Résumés dusted, updated, sent off, lunches and coffees with old friends. Eventually those efforts waned. Determination seemed to evaporate. Perhaps finding a suitable job at age fifty-five seemed hopeless, perhaps the initial lack of response signaled that it wasn't worth trying. From my twenty-two-year-old vantage, it didn't register as hopeless in my father. It looked like settling in, and it was strange and frightening to see. If shame and fear existed— motivators both—they'd been pitched into a deep hole in the dirt far below the frost line and buried there under dark earth so that any evidence of either was darkly gone, worm-chewed and composting, inaccessible even to him.

"What are you up to today, Dad?" we'd ask.

"I've got a dentist appointment at three," he'd say, and we'd wait for more. With nothing else to do, a teeth cleaning was a day's duty. He put on a suit for his appointment, carried his leather briefcase with him, let the world know he was a man who wore an eighty-five-dollar tie. At the time, it struck me as a lie. He tricked other people in the waiting room, crinkling through dated issues of *Time*. He tricked the hygienist and the dentist as they flossed him, had him rinse and spit. He tricked the driver stopped at the light next to him on the drive back from the appointment. I knew the truth, or thought I did, with the indignation and confidence of someone who'd been paying her own rent for less than a year. The tie, the briefcase, these were dishonesties meant to make people believe that this successful man of business was headed straight back to the office after his teeth cleaning.

There's so much I didn't know then. It wasn't until after I left my job at the newspaper that I realized how significant a part of my identity working there had become; it was how I understood myself and made myself understood to other people. *What am I now?* I wondered, jobless. And that was after less than a decade. My dad's working life had spanned more than thirty years—that's a hard habit, and a hard self-conception, to break overnight. I couldn't imagine the terror of losing your job (and part of your understanding of yourself) at that age. No wonder he did what he could to maintain it. No wonder he put on a suit and carried his briefcase and selected from the hanger a favorite tie. Maybe it wasn't a lie after all, not a trick, but a continuation of how he understood himself, and wanted to be understood.

Months passed. My father still wasn't working. My mother said quietly to me over the phone: "He's always *around*, he never leaves the house." She started waking up at four-thirty in the morning to have time to herself in the house alone before going to work. Picturing her in the predawn darkness, showered and dressed, alone with her coffee, puts a sadness in me I don't have words for. "It's the best part of the day," she said.

Isn't this a problem? I worried. *Shouldn't he be working?* I asked my mother: *Do you ever say, hey, hi there, you need to work?* She didn't want to nag him, she told me. She said if she asked him once, she wouldn't be able to help herself but ask everyday. This seemed like an error to me. Even with what little I knew then about relationships, joblessness, I knew enough to sense that this was the wrong approach. Why not summon enough willpower to ask once, and then ask again sometime later? Not nagging seemed like a good impulse, but doesn't someone who spends the hours of his day shifting backgammon pieces across a board on a screen deserve a bit of a nudge? And what if you exchanged the word *nag* with its connotations of henpeckery and the worst of tiresome wifehood, with a word like *challenge*, or even the neutral *ask*? She never asked him how his job hunt was going, or if he was having any luck, or if he'd considered this or that. There was neither encouragement nor pressure. Neither *I'm rooting for you* nor *Figure your shit out fast*. He took her silence as a lack of interest, and his voice was pinched with blame when he talked to me about how *if only Mom had asked me about it*.

At age twenty-two I recognized, with the deep sadness that accompanies seeing your parents as fallible, that this was a failure. Of course my mother should've asked him and it should have had a note of urgency. And of course my father should have been able to motivate himself to look for work regardless of what my mom was or was not asking.

He went through a scone-making period. A carton of buttermilk, thick and sour smelling in its cardboard container, was a constant for a time in the door of the fridge. And in the morning, a batch of warm dough triangles—ginger lemon, blueberry, raisin, orange zest—cooled on a cookie sheet. Besides the batch just pulled from the oven, on the counter three dozen more from previous batches were heaped like throw pillows in plastic zip-lock bags. They were good to eat. Not the dry, bland saliva-suckers that bad scones can be, but flaky, gentle-flavored, delicate, and hearty. There were just so many of them. And I can see how making a batch of scones could feel more productive than sixty hours a week at an empty job. The results are quantifiable, concrete in the world, useful. Dough is an essential thing. But is it enough?

Perhaps for him, for a time, it was. "We're ensconced," my brother joked, gesturing at the bags on the counter beneath the kitchen window.

That winter in Cambridge, waiting for Mary to call, when I wasn't working and didn't know when work would come again, I felt a lack of purpose. I wanted to be useful, to contribute to the world. Building a set of bookshelves or laying down a floor wouldn't reverse global warming, but it was better than what I was doing, heaping myself on the couch, wrapped in a blanket, staring blankly and absorbing nothing while the opinions and pictures, the news and noise rolled slowly up the computer screen and disappeared over the top edge, as though it all tumbled off some invisible cliff behind the machine and fell into the nothing. The hours went with it.

My father refused to consider taking jobs he felt were beneath him, even in lines of work that interested him. A life-long and dedicated reader, he scoffed at the idea of working in a bookstore. A serious gardener, he scoffed at the idea of getting a job at a nursery. Such suggestions, offered up by my brothers and me, were insults to him. Pride kept him from recognizing the value in being part of something bigger than himself, that there was honor in helping someone find just the right book or in showing someone how to make their hydrangeas thrive. He was too good for work like that. So instead, he did nothing.

When tomorrow is always an option, an empty place, the rush to get things done does not press in. Things go undone and undone, days curling off and falling to the floor like wood curls chiseled off a piece of useless pine, swept up into the dustpan and thrown away.

Visiting home one weekend, I passed through the office where my father looked at the Internet. He said something about bluefish. I went on my way, confused by the terrible new mix of feelings this provoked, pity and frustration, disappointment and an awful ache. Later that afternoon, I saw my father out the window in the backyard with his fly-fishing rod, standing on the grass as dusk cast a blush on the sky. He stood, rod across his right shoulder, and with a quick flip of the wrist, practiced and precise, whipped the rod through the air, right arm flung out in front, aiming his lure at some imaginary river. Again and again, he flicked and flung. I watched as my dad cast on the grass toward a wooden fence, heard the whistle of the rod as it whipped through the early evening air.

I wondered if he ever went fishing.

Sometime in those months, which stretched into years, my mother came home from teaching children at the preschool she directed and my father came downstairs from the computer and into the kitchen and said to her, "We're out of milk." My parents got divorced.

M y sleep became increasingly fragile. Hours of fret dissolved into fluttery sleep. What will I do tomorrow? How will I pay rent? What are the lead paint, asbestos, wood-dust particles doing inside me? I had time and I looked *in*, peered down inside myself, and what I saw was nothing. The darkness was total and I wondered, in there, where I was.

I had saved money diligently when I worked at the newspaper, living cheaply and amassing what I considered a small fortune. But I was nearing the end of it. My confidence dipped with the number on my bank statements. How is this where I ended up?

And I missed the work. I missed Mary. I missed the way a one-by-four plank of Brazilian walnut felt in my hands. I missed coming home, muscles weary, hungry and dirty. I missed the feeling of a fatigue well earned. I missed the satisfaction of something being further along at the end of the day than it was at the start. I wanted to go back to making things. I wanted to make a living.

Question marks and idle time framed the wall of a persistent anxiety, of self-doubt, of true slump. My savings were almost gone when I heard through the grapevine that my old newspaper had a staff-writer position open. To go back, to return to the place I left—the thought of it brought a tight feeling, a quiet shame. But the pull of a paycheck, health insurance, and the feeling of being part of something again was a tide stronger than the shame. And maybe I could still do some carpentry on the side. I called my old boss, feeling defeated and sad, and told him, *Yes, I'm really so interested. I'd love to come back.* And he said, with some teasing satisfaction in his voice, "The carpentry work didn't do it for you?" I was standing on a street corner in Beacon Hill, next to a luxury SUV the size of a hippo, watching strollers and fancy moms and small dogs slide up and down Charles Street. I clenched my jaw when he said it, closed my eyes. Carpentry work *was* doing it for me. Hard, dirty, and possibly toxic at times, it was work I'd come to love; it was just that there wasn't enough work. Standing on that sidewalk, phone to my ear, I realized how much I'd rather pull up carpet in a damp basement than return to where I'd left. But the urgent need of cash and health insurance, the urgency to fill up my days with purpose, outweighed my pride. "I'd like to come back," I said.

"Well, we'd love to have you back," said my old boss. He told me they'd have things sorted in a few days.

I shivered on the sidewalk. Down a narrow alley barely wide enough for a single car, lanterns lit up, fake flames for streetlights, and the effect was back-in-time, lamplight flickering off brick, and couldn't it just be 1850? The mothers in their pearls and the baby carriages and the small dogs passed by. The cold air made my eyes water.

On the walk home toward Cambridge, I tried to convince myself that this was what I wanted. I made myself remember the shitty jobs and the times I hated carpentry work. Think about working at the Haitian sisters' house, I told myself, where they kept the heat at eighty-seven, and one of their sons had a stomach flu and there was puke on the floor in the hall and all over the bathroom sink. Think about how down in their basement, before we installed the bamboo floor, we had to pull up dirty old carpet, and how the basement was as warm as the rest of the house, and the carpet gave the room a moldy stink. Think about how Mary worked on making new stair treads for the basement while you ripped up the rug, and how you envied her, with her tape measure and her wood, as you yanked and tugged on that mustard-colored carpet. Think about the bits of grit and dust and how they seemed to have a magnetic attraction to your face, and how sweat slid down your temples as you used a blade to carve through the shag and base of the carpet, like skinning a Muppet. Remember the way the staples chewed at your arms like a thousand demon bites. Remember how much you hated that day.

And I tried to think fond thoughts about the newspaper I'd left behind—that people I liked and respected still worked there, a few of them anyway, that I'd be writing all the time. I'd get to walk to work again over the smoot-measured Mass Ave Bridge. (This, of all things, held the most appeal.) I thought of the regular paycheck, of being able to put money back in my savings account. Yes, this was the right path, it's okay to go back. It's no failure. (*Failure*, my brain kept saying.)

Two weeks passed with no word back from my old boss. Then I heard that they'd hired someone else.

In my dark and muddled state, the disappointment and anger came not from getting passed over for the job, but from not being told. It wasn't so much that I wasn't good enough for it—it was worse: I'd been forgotten.

A wake at night, to steer my mind away from the thoughts that clamored louder in the dark, the ones that sent my heart pounding, the ones that clawed, I thought about a house I would build for myself. I'd lie in bed and lay out floor plans, erect walls, design tile floors and book-shelves and bedroom windows, kitchen cabinets, counters, and pantries, consider thresholds, light, and warmth. I'd been in so many homes with Mary, and I took what I knew from them. I'd flip through the jobs in my head, the ones I'd liked and not. I'd start at the frame, see the skeleton, see the joists running across the floor. I'd cover them with plywood, subfloor, hardwood, cherry maybe, or rustic reclaimed wide old pine. I'd feel the mallet in my hand and imagine the board-by-board effort. I pictured the pneumatic floor nailer, powered by compressed air, and the thump-pop of striking the rubber button on the nailer with the rubber-headed mallet. The nail clunking through the wood is as satisfying a sound and action as there is. Putting floors in gratifies. A room is transformed when it goes from cement or subfloor—guts, unfinished and dirty—to smooth finished wood with knots and swirls that lap like waves. Color will deepen with time. Scratches and dents will scar the wood, life will make its mark, wear on it.

In my mind, at night, I'd position walls and windows, frame them first with two-by-fours, hammer and nails, with windows across a whole wide wall, thick header atop to spread the weight of the wall. After the framing, drywall, taping, mudding, and paint. I'd see where the door to the kitchen would go, cut pieces of trim in my mind, fill the nail holes with putty. I'd picture a fireplace and figure I'd have to hire a mason. I put a skylight over the tub on the second floor like I'd seen at a place where we'd built big built-in bookshelves. I imagined moving from room to room to find the right flow. I wanted rooms, doorways and walls, space divided by function, not a kitchen that bled into a dining room that bled into a living room. I'd make the porch in the back out of Brazilian walnut and watch it fade from its syrupy cinnamon color to worn gray like the clapboard shingles of houses on the coast.

Thinking this way bathed my brain in calm late at night.

In her essay "On Coming Home," Joan Didion writes about

returning to her childhood home. "Paralyzed by the neurotic lassitude engendered by meeting one's past at every turn, around every corner, inside every cupboard, I go aimlessly from room to room. I decide to meet it head-on and clean out a drawer, and I spread the contents on the bed. A bathing suit I wore the summer I was seventeen. A letter of rejection from *The Nation*, an aerial photograph of the site for a shopping center my father did not build in 1954."

When my parents split, my mother took an apartment in a small town in mid-coast Maine. She has all the photo albums. When I visit, I look through them after she goes to bed, like Didion cleaning out an old drawer.

My father rented in a town on the southern coast of Massachusetts. He'd put all his possessions in storage. The house he lived in was furnished. Maybe it felt like home to him after a time, using a stranger's spoons. It never did to me.

My grandmother's house became the one place that did feel like home, that held within its walls all the stages of my life, the whole show. And not just mine, but the lives of my whole family—my mother, and my parents together, and my brothers, and, to a lesser extent, my cousins, aunts, and uncles, a family extending and extending, everyone bound up in this one place. Whenever I visited, I wandered room to room. I opened all the drawers, seeking treasures, memory triggers, connections. A framed photograph tucked at the back of a desk shelf shows my mother and her four siblings and their spouses, all just starting out with their own lives, before kids came. They're gathered for Christmas with my grandmother, the plates of a big meal haven't been cleared off the table. My grandmother is in the middle of the photograph, looking slimmer, taller, and tense. Were the photograph to be taken now, four out of the five spouses would be absent: one dead, three divorced. Blood works as a kind of clamp. It presses us together, erasing distance, even when so badly we want it, with a claustrophobic feeling—too close, too close—and that visceral constriction of hunger, a squeezing sort of lack. It binds us so no matter what, we always share an edge.

I went to my grandmother's house in March of that long winter. She no longer lived there—mind fogged and body strong, she lived in an assisted-living place in Bedford, Mass. The house remained in the family, and by that point the walls of the apartment had gotten too familiar, a distorting sort of closing in, strong as a vise and gaining strength with each day. The word *escape* came to mind, not in the tropical vacation sense, but of the prison break. My grandmother's house was my favorite place, and I'd gone past the point of not being able to justify something good for myself—for a while I'd felt I didn't deserve a break, a change of scene, pleasure. That had given way to a desperation, a knowing of the necessity of being away, even for a few days.

The boards in the attic at my grandmother's house are wide. Yellow pine, an inch thick, eight and ten feet long, nearly two feet wide. I slept up there in the summers as a kid. Cobwebs dropped with dust from the beams, and the creaks in the night, the sighing and squeaking of wood under weight, those sounds are ghosts. Benevolent ones, but ghosts. A stairway leads up to a hatch in the roof, a heavy trap door out into the sky. I spent hours up there in the evenings as a teenager. The attic smelled like wood and dust, something dry and old and pressing in with its aliveness, a different sort of press than the walls of my apartment.

Looking at the boards that line the walls and the high peaked roof, the saying *They don't make them like this anymore* applies, because they cannot make them like this anymore. For boards this wide you need trees this wide, and we've mostly cut down those trees and the new ones haven't had time to thicken, a ring added with every year. (Imagine if each birthday we were marked in some way, physically scarred, not by creases by the eyes, or softening flesh, but by something you could count and tally.)

Old photographs of people I never knew the names of, long dead, filled frames. There's an accumulation of old quilts, suitcases with rusting latches, a trundle bed, busted chairs moved up here from lower parts of the house. It was best in rainstorms, when the rain pattered right on the roof above, a wet and steady hymn of rain on roof like the sound of small wings beating. The house gave loud voice to the wind. I came to know what direction it was blowing based on the pitch of the howl or wail or whisper.

In a corner of the attic, in a dusty mess of boxes and old suitcases, leaned a section of board just under an inch thick and about three and a half feet long. Sixteen inches wide, this was old growth from a wide tree from a hundred and fifty years ago or more. Scratched slashes marked one side, the slash count for six. I took it with me when I left.

It brought comments as I walked through town with it under my arm.

"You going surfing?" one man joked.

"Now there's a girl who's really *bored*," punned another guy.

"What type of wood is that?" asked an older man with a mustache.

"Pine," answered his companion before I could answer. He was eagle-nosed, with white strands of hair running over a smoothed dome. He was dressed for the weather. I asked if he knew what the markings were about.

"Yes," he said, and he explained that a lot of the old houses were moved from one place to another, and this was a way for the builders to know which pieces went where when they were putting it back together again.

I took the board home and spent some hours with sandpaper, grinding away the dust and dirt, smoothing out the rough surface. Dust drifted on the breeze and it smelled like the attic. It burned my nostrils as rough gave way to something light and smooth beneath. Rings and waves revealed themselves. Hidden under the rough cut of a crude saw from more than a century ago, dark knots got darker, eyes in the wood. Swirls and waves like ridges on a sandbar seemed to rise from the surface as the sandpaper abraded layers away. Rubbed and rubbed, the wood shifted from dusty dark red-brown to something paler, the whirls and rings and the lines of the grain a salmon color, a living pinkish. After hours with the sandpaper, finer and finer grade, to rub my palm across the wood was to touch velvet, baby skin, a cheek. This transformation surprises me every time, the way wood, under a bit of effort, can be so soft to touch. It's a miracle of transformation, and it thrills me. How changed it is, and still the same. I smoothed a finish over the wood, a combination of polyurethane and tung oil and linseed oil. It smelled like apple cider, a vinegar sting, with the sharp and singeing smell of turpentine. I smoothed it on the wood with a torn piece of a soft pink tank top I used to wear. The finish looked like honey from the jar. I smoothed it on and the wood drank it in. The color shifted again. The brown-reds re-emerged as though answering a magnetic draw from above, from pale to rich, the color of fall. The eyes, big knots, got black, all pupil, and the swirls and rings became a rich dark orange like a flame against the base color of the wood. I attached iron hairpin legs and it is a table now, this slab of wood from the attic in my grandmother's house. The slash markings are on the bottom, a secret, a reminder that everything, almost everything, can be put back together again.

I n early April I grabbed my phone, resolved to call Mary again, to be firm and tell her if there wasn't any work in the foreseeable future, I needed to find another job. I faltered as I scrolled to find her number. It was not a call I wanted to make. The possible answer—*nope, sorry, nothing doing, you're on your own*—was one I did not want to face. I stared at her name. To summon the courage, I imagined telling her about the table I'd made, and that felt like a good enough reason to call. I pressed her name on the phone.

"How the hell are ya?" Mary asked without saying hello.

"Ah, okay, okay. It's been so long —"

"I was just going to call you. Got a job in the South End starting Monday. Could be a weird one. Does eight-thirty work for you?"

Yes, I told her. Yes. Eight-thirty worked for me.

Chapter 5

—

SAW

On severing a part from the whole

That Monday morning, we arrived at the place on a tree-lined side street of handsome brick row houses in Boston's South End, a high-rent neighborhood with a long list of good restaurants and galleries. The condo was a couple blocks removed from the action of Tremont Street, the Boston Center for the Arts, its Cyclorama space and art studios, the Boston Ballet, an oyster joint, a subterranean bar that aims at a bohemian feel, and boutiques of expensive bibelots. The front door was locked and Mary didn't yet have a key.

She rang the bell. No answer. She rang again. "This could be a problem," she said. We were bashful around each other, not having seen each other in almost six months. She pulled out her phone and called the woman. No answer. "We're on your front steps, hoping to get in, see you soon," she said to voicemail.

We stood on the stoop a few more minutes. The morning was cool in a way that gave hint of coming warmth. Winter, which lasts and lasts in Massachusetts, had broken. Buds and greening trees hadn't started yet, but it wouldn't be long. A softness in the air whispered at the bloom to come. I pressed the doorbell. We could hear it buzzing up on the second floor. Finally, we heard footsteps down the stairs and the door opened. The woman stood there, hair rumpled, puffed-up pouches below her eyes, in loose pj bottoms and a T-shirt that fell off her shoulder. "Hi, Nidhi," Mary said. "Sorry to wake you."

Without a word, Nidhi turned and climbed the stairs. We followed. She entered her apartment and took a right down the hall. "I'm still sleeping," she said as she closed her bedroom door.

We'd been hired by a real estate agent Mary knew to fix up her condo, to repair cabinets and banisters, level slanting built-in shelves, paint over some half-done paint jobs ("I'm thinking *manic episode*," Mary said of one section of bright purple paint), redo the bathroom with elbow grease and a new vanity. In general, to gussy up the place before it went on the market. It needed gussying. And it was a good first job to return to—a lot of quick fixes and easy items to scratch off the to-do list, nothing too challenging while we got our feet back under us.

"You should've seen it last week," Mary whispered. "There was crap everywhere. You couldn't move." I felt the closeness of being in on something together; after months apart, all it took to get that camaraderie back was a *listen to this* as we lugged and unloaded. Mary filled me in a little more. The woman was working with a professional organizer, and though they'd made some progress, navigating through the place was a challenge. Lamps here, crates there, trash-bag heaps, boxes marked "!!!fragile!!!," another box labeled "lingerie rarely used," hair-spray cans, headbands, a basket full of sunglasses, mirrors—multiple mirrors, large ones—leaning at precarious angles, making elongated fun-house reflections. The morning was bright and clear; inside it was dim like the moment before evening gives way to night. The shades were drawn, and over them hung thick, dark towels. Underneath, taped against the windowpane, were old record covers. An early R.E.M. album showed Michael Stipe with glasses and thick wavy hair. The air felt as though it'd been stuck in there, unchanged, for weeks.

"I don't know if I'd call her a hoarder," Mary whispered as we climbed the stairs with tool buckets, drop cloths, and a couple gallons of paint. But I bet that's because the hoarders we see on TV have cat carcasses rotting under piles of cracked china and fabric swatches and old dolls. The place didn't horrify, but it was evident that this was more than the chaos of an impending move.

"I started smoking again," Nidhi said when she emerged from her room around noon, still in pj's, hair no longer a storm around her head. "Because of the move. I quit ten years ago." She was moving to Pennsylvania, to be closer to her mom. "I've lived in this place thirteen years," she said. "Time to spread my wings." It sounded practiced, the way she said it, as though she was still trying to convince herself, or repeating someone else's words.

I felt clumsy with the tools, out of practice, and my heart beat hard in my chest and my breath came fast carrying saws and tool buckets up the stairs. But it felt familiar, too, like returning to a recipe you haven't made for a while—you remember the knife and when to add the salt, but there is hesitation, a stuttered second-guessing, *is this right, does this come next?* Digging through the bucket for a screwdriver, chopping a piece of two-by-four with the clean smell of pine filling the room, I thought, *Of course. I remember this.* And I got a small smile on my face. I also got a splinter.

Mary had me try to deal with some collapsing bookcases in the bedroom. Tacked to the front of them was a list of questions. In the neat and optimistic handwriting that would belong to someone who helped people organize for a living, it listed criteria for sorting and keeping:

Is it current? Is it of good quality, accurate, or reliable?
Have I worn it in the last year?
Can I wear it with at least three outfits that I have?
How many items like this do I already have? Choose the best and
    limit the number I keep.
Is it unique?
Does it represent who I am and do I feel comfortable in it?
Does it fit in a flattering way?

I imagined Nidhi picking up a necklace with a broken clasp, an old shower caddy, a thrift-store cardigan with buttons loose, and reading through the list. Keep or toss. Treasure or trash. Stay or go.

It brought to mind an image in *Moving Out*, a book of photographs by Robert Frank. A mostly blurred black-and-white picture shows a rock with some snow on it in the foreground. Two telephone poles flank the shot; fuzzed power lines stretch across the horizon. The white of the snow on the rock is a sharp glowy shape against the middle gray of the rest of the image. It's less a landscape and more an abstract, a mood. The atmosphere is grim, like a muted Sunday afternoon in February when it seems like winter will go on forever. Scrawled on the print with a brush or a fingertip are the words *HOLD STILL Keep going.* The letters look as though they were written in blood.

"The sunlight is jarring," Nidhi said.

Mary was gentle. "We've got to take the towels and tapestries down to paint."

"Mornings are difficult for me. You might've noticed." She laughed. She was likable, a little nervous, and gracious. She thanked us for making her home look better. She offered opinions on paint color, said she wasn't sure about the beige in the hall.

"I should just stay here now."

"She's got to start thinking of this as not her home anymore," Mary said later as we sat on the front steps eating lunch.

She and I hadn't found our rhythm yet, were still easing back in. It felt like moving through a once familiar house at night, a vague sense of where the sofa was, and how to avoid knocking against the corner of the table there, but a bumbling feeling too, a hand out in front to make sure you didn't walk into a wall.

"Dusting off the cobwebs," Mary said. "We got to get your muscles back."

In Nidhi's kitchen, a sticky black grime coated the counter by the fridge. Mouse droppings dotted pans on the stove and piled up in the dark corners on the counters. Open soda cans, half-drunk, filled a cabinet. The face of a small drawer next to the oven hung loose, its handle dangling off it. When I opened the drawer to fix it, I saw fifteen, twenty orange prescription bottles, some pill-filled, others close to empty. Mary was up on the ladder, leveling upper cabinet doors above the fridge. I looked up at her.

"I know," she said. "Don't look."

I didn't want to look. Of course I wanted to look. I wanted some clues as to what this woman suffered. What were these pills—so many pills—used to fix?

I fixed the drawer. It closed smooth, right on its rails, with face attached and handle screwed tight.

Spending time in other people's homes was one of the best pleasures of the carpentry work, and I felt especially grateful for it now after so long inside my own apartment. To see what cereal other people ate, how they brewed their coffee, what pictures hung on their walls, what books filled their shelves. The bookshelves always drew me first. Whenever Mary was outside having a smoke, and sometimes when I should've been installing a threshold, I'd look to see what lined the shelves in people's homes. And I'd look at whatever was open on the desk—Post-it notes with phone numbers, a photograph of the couple looking younger, an obituary clipped from the newspaper. "Stop snooping," Mary would say. Are there cats? Kids? Is the bed made? Would I want to live here? Would I want to live like this?

Is it every human's impulse to peer into other people's windows? What a small specific pleasure it is, to see someone in a moment of their living, a glimpse of someone standing at the

stove over a steaming pan, pulling sheet corners over the mat-
tress, brushing teeth, taking off a sweater. Carpentry allowed
this glimpse into other people's lives, not in fast glances through
lit windows, but through the front door and into their rooms.

Taped to the frame of the door into the kitchen at Nidhi's
place, a hot-pink Post-it note read: "How can you be so
judgmental?"

It was a judgment itself, self-turned. An unkind reminder:
what gives you the right, who do you think you are? And it
made me nervous, too, that just as we were in her private space
and coming to know her, she was watching us. If you need to
write a note to yourself reminding yourself not to be so judgmental, chances are you haven't broken the habit yet. It made
me more aware of how I was with her, how Mary was with her.
Working in someone else's home, one enters into the private
space of a stranger, and a strange intimacy occurs.

Nidhi caught me looking at a photograph on her fridge: a
beautiful woman with bright eyes and thick hair sitting on the
railing of a deck by a hedge. The woman in the picture wasn't
smiling, but she looked happy and the light looked like almost
evening light.

"That's my mother. Isn't she beautiful?"

"She really is."

"She never seems to get old. My dad looks young, too. He
jogs eight miles a day. I've got good genes. Guess how old I
am?"

I suspected late thirties, but feared her puffed eyes and
tired mouth were the result of the drawer of pills, that they
had aged her beyond her years. I took a few years off.

"Thirty-three?"

"Ha!" she laughed. "Ha, I told you. Good genes. I'm forty-four." She seemed well pleased, and I got the feeling that this was a game she played with a lot of people.

In the hall that led to the bedroom, scrawled on the wall in crayon, letters eight inches high: "6 hours earlier in Lulu!!!" Her sister lived in Hawaii, she mentioned at some point. Did she keep calling her too early in the morning?

We painted over it on a Tuesday.

A week or so after we finished up there, I thought I saw Nidhi on the street, wearing huge sunglasses and walking a big black dog. It felt odd to see her out of the context of her home, and I felt too nervous to say hello. I looked at my feet and crossed the road. I don't know if she saw me. I don't know if it was her.

So started a season that rocketed along. Later that summer, we were hired to do a total renovation of a kitchen. Nothing would go unchanged in this lovely third-floor Cambridge condo kitchen. New floors and cabinets. New countertops. New appliances. A doorway would be moved. A pantry would be built. The stove and sink were shifting from one side of the room to the other, which meant pipes had to be realigned. It was a big job, and Mary's excitement was contagious.

Two women in their early fifties owned the place, Alice and Bettina. Bettina, from somewhere in the Black Forest, was large in a big-boned Teutonic way, and spoke with a gentle German accent. She'd tilt her chin toward her chest when speaking, which gave the effect, despite her height, that she was looking up at you. She gave an impression of forgiveness, which softened her imposing presence and likely suited her students at the university where she taught. Alice was large in a short, round way and her thick breasts hung braless like sacks of coins. The kitchen, it was clear from the start, was hers. She had designed it, and she would be the one to grill meat on the restaurant-grade stove and to roll delicate pastry dough on the marble counter. She also worked from home, so we'd be seeing a lot of her.

A sick heat had settled on the city, and it was only getting hotter. Boston in July is a soup of swelter and summer funk. Thick air makes every hug, every article of clothing, a torture. I pictured temperature degrees as invisible rods in the air, dense packed, heavy with moisture, settling on the skin, pressing as you moved. I sweat and sweat, heat-stunned and dulled.

On our first day at Alice and Bettina's we unloaded the van and carried the tools up to the third floor. Our steps were fast and the loads felt light, carried with the excitement and optimism of a new gig.

"You should see the tiles Alice ordered," Mary said. She rubbed her fingers together in a way that said *pricey.* "They're gorgeous. I always tell people, when they're trying to design a new kitchen, that they should pick one thing they want to splurge on. Cabs, tiles, new island, whatever."

Tiling had been the first thing I'd done with Mary, and it was always one of my favorite parts of a job. The variety appealed—each kind of tile had its own personality and place. Tiny white coins make a good match for the floor of a small bathroom. A grand, high-ceilinged front hall can accommodate massive panels of tile for high heels to echo off of. A kitchen counter tiled with lapis blue brings warmth to the room. Texture varied: shiny smooth, earthen matte, rippled and gently ridged. Color varied: sunset terra-cotta, beach-stone slate, the promise of clarity and clean living of plain pure white.

Once the tools were upstairs, we surveyed the room. The demo had already been done so the room was blank, stripped of appliances, cabinets, and floor. The fridge was the only thing left in there; we'd need to move it before we started. Mary gave a quick order-of-events rundown and talked through where things would go. Fridge on the wall to the right; sink to its left; oven facing the fridge from the peninsula in the middle of the room, which would jut out between the two windows on the wall opposite us. There would be short cabinet corridors on either side of it. Open shelving would go on the left wall, a slab of marble countertop below the window on the left, and the pantry in an area by the door out to the back deck. I nodded as Mary talked, trying hard to position everything in its right place. It takes practice and imagination to conjure up a full, functioning room out of a blank one. Staring at this emptiness, it seemed near impossible that this would be an actual kitchen again. But I could feel the potential, too.

"Let's get the fridge out of the way first," Mary said.

I reached around the door of the fridge to get a good grip on it, and accidentally pulled it open. All at once, a potent, terrible smell knocked us both backwards. The sour stench of spoiled milk mixed with the musty stink of raw rot, a stale plastic smell as though electricity itself had decomposed. Mold dusted and slimed all over the shelves and drawers, a creeping black fuzz. Alice and Bettina had gone to Germany for a few weeks to avoid some of the upheaval of the renovation and had unplugged the fridge before they'd left. But they'd forgotten two tubs of yogurt and a block of Emmentaler cheese, and temperatures had hovered around eighty-five degrees since they'd left a few days before. Mary pulled open a drawer to find some now-unidentifiable plant matter, a mucusy vegetal slurry.

So instead of a quick shift of the fridge and getting on with the framing of the new doorway, Mary and I spent an hour scrubbing every surface of the fridge. "Take the drawers to the tub," she told me. I washed mold off the plastic with water and Lysol and a green sponge and thought about the stutter steps of a new project. When I tell people I work for a carpenter, no doubt they envision pale curls of wood, the homey Christmas smell of pine, the quiet contemplation of craftwork. But here I was scrubbing mold off a refrigerator's veggie drawer in a stranger's bathtub. Often, at the start of the job, the work consists of things we don't expect, that have little to do with a carpenter's training or expertise.

Isn't this often the case elsewhere, too? When we picture the lives of other people, we imagine the most exciting parts, the ones rich with drama and *living*. The ER surgeon reattaching a man's leg after a car accident. The painter finishing off a portrait, paint on her wrist, and getting into bed with her subject. The farmer trundling in from a day of harvest, tossing a dirty sack of fresh carrots or onions on the table. Imagination is the enemy sometimes, in how fully we can bring to life the passion our current love shared with someone else before, in how fire-filled someone else's existence is compared to our own.

But of course, most of us spend our time figuring out what to make for dinner, trying to remember to buy another roll of paper towels. Our romanticizing is perhaps an act of hope, that those sorts of lives are possible to live, that it's possible to find challenge and satisfaction in our work, to have our bodies lit up with lust, to happen upon those conversations that go deep into the night when voices get quieter and truer things get said. In our imaginings of other people's experiences exists an ambition to exist in our own in the fullest way. *You're a carpenter, it must be amazing to make things!* And it is. Except when it isn't.

T hree days into the job and we were ready at last to start on the floor. I'd gotten a call from Mary the night before. "So listen, I've got a bunch of running around to do tomorrow. The tiles are getting delivered between nine and ten. They're going to leave them at the bottom of the stairs. If you just want to be there for the delivery and bring them up, we can call it a day and I'll see you back there on Thursday."

Bring a few boxes of tile upstairs and call it a day? Great.

I sat on Alice and Bettina's front porch and waited for the tile guy. A few minutes before ten, he pulled up in his truck, shoulders like cantaloupes and forehead dripping with sweat.

"Hot enough for you?" he said. He began unloading boxes, two at a time, to the landing at the base of the stairs. "You got some help with these, hon? Or are you bringing all of them upstairs yourself?"

"Just me."

"This building got an elevator?"

"No."

"You got your work cut out for you today. You stay cool."
He climbed back into his truck and roared off toward another
delivery.

Who needs a fucking elevator?

Twenty-five boxes of these tiles plus two sixty-pound bags
of cement sat at the base of the stairs like the beginnings of a
fortress. Mary had been right about the tile. Beautiful five-by-
five-inch squares, slate gray like smoothed stones and no two
alike. Some had bumps and small pits, some had striations of
paler gray like a good luck ring around a rock at the beach.
Even in the boxes, it was easy to imagine them underneath
bare feet as you stood by the stove scrambling eggs on Sunday
morning or tip-toed through in the evening to pour a glass of
water before bed. I looked at the stacks. Fine. A lot of trips,
but I could probably take two boxes at a time like the tile guy.

I lifted a box—*oh, shit*—and I put the box down. Then I
laughed. I would not be carrying two boxes at a time. I couldn't
believe the weight. A box the size of a loaf of bread, and each
one weighed thirty-five pounds. That's like holding more than
four gallons of water, the same as a sack of about twenty-eight-
hundred quarters. Thirty stairs rose between me and the third
floor. By ten that morning, it was already edging up over
eighty-five degrees.

I got myself into mule mode. One box at a time, up and up,
steady steps, then barreling down the stairs for another load.
Box in my arms, up we go, then bouncing down. It was hyp-
notic. I was a body moving up and down, brainless and physical.
All I needed was muscle, patience, and the will to get a thing
done. It was similar to the task of cleaning the moldy drawers:
boring, necessary, underimagined.

The pile of boxes dwindled at the bottom of the stairs and grew at the top. Ten boxes left, then four, then one, and I realized I should not have left the two bags of cement for last. I climbed eight hundred and ten stairs that day, hauled up nine hundred ninety-five pounds, nearly half a ton. The feeling that resulted from the effort, the satisfaction, was so different from the one I knew putting a final period on a book review or a profile on deadline.

Finishing a piece of writing, the sensation was relief coupled with a spentness, a short temper and depletion, grinchy and hollow. After a deadline, I experienced a pinched feeling behind the eyes, and the next person I'd encounter would get strained smiles and diverted, unfocused attention. The more fully I existed in the world of the writing, the more removed I'd feel from the world as it existed around me, and the transition back, particularly after rare moments of writing flight—when the words come and there is nothing else—would grind. Almost immediately upon finishing a piece of writing, the glow faded, and all I'd see were the flaws.

Work with Mary was different. I looked back on everything we'd built with satisfaction and pride, even the things that didn't deserve it. The bookshelves for a rich psychiatrist with a grand piano: without question the best bookshelves that have ever been built. That bamboo floor we installed in a basement to turn it to a bedroom: who cares if the floor was the color of Band-Aids: there has never been a better bamboo floor than that. The deck stairs in Somerville: I could run up and down those stairs for hours, they are exactly what stairs should be.

Lifting and hauling those boxes of tile, I couldn't ignore my own sweating and panting, my muscles flexed and straining.

When I placed that last box and final cement bag by the kitchen door, I felt buoyant. My whole self felt more honest, more useful, and more used. There was no grinding back to a different world. I'd been there the whole time. I took off my shirt and wrung the sweat out over the bathroom sink.

I called Mary. "Finished," I said.

"Wooohooo! You must be sweating, girl. Drink some water and try to find someplace to swim this afternoon. I'll see you bright and early tomorrow."

I skipped down the stairs and locked the door behind me. The view felt longer leaving work that day. The air was thick as I walked home, and people's foreheads were damp with sweat. Dress shirts clung to backs and chests, and the leaves on the trees seemed a more saturated green, benevolent somehow, as though aware of the heat and eager to shade. I smiled at someone across the street and he smiled back. Everything was okay, everything would be okay, the small snarls and woes were just that—they evaporated against a much bigger, much stronger tide of connection to life. Walls come down, the ones that block our view of each other and the leaves and the sky, that divide us from the awareness of being alive.

And those boxes were fucking heavy and I was glad to be done with them.

At that point though, I'd graduated from being just a lugger. Instead of just helping Mary with what she was doing, or watching her do what she was doing, I was on my own for certain parts of projects.

Mary had me build simple birch plywood cabinets to be

tucked away in the pantry, a small zone that would serve as a transition from the kitchen to the back porch. Mary and I ripped sheets of three-quarter-inch eight-by-four plywood down to size on the table saw (a rip cut is one made parallel to the grain). I chopped the sides and tops of the cases with the miter saw and attached them together with wood glue and a nail gun, the flinty smell rising after each shot. I set them on the ground so they looked like high walls to a big sandbox and fastened a piece of quarter-inch plywood to the backs of the boxes to keep them from wobbling. The backing piece braced them. If you cut the front and back panels from a box of popsicles, imagine the movement if you then put your hands on the remaining edges and shifted your hands up and down. The same thing happens with the cabinet boxes; the back panels stabilize the shifting.

The boxes and shelves needed trim to cover the ugly unfinished look of the plywood, which is made of thin sheets of wood glued together cross-grained—the grain of each sheet alternates direction with the sheet before it, which makes it resilient against bending, swelling, shrinking, and splitting. It's stronger than wood you find in nature and much less expensive than solid wood. Alice had splurged on tile, saved dough on the pantry. For these cabs, which no one would see the sides of, plywood was just the thing to use.

To cover up the plywood edges, I measured, marked, and cut pieces of one-by-three-inch poplar trim to line the cases, and one-by-two for the shelves. Poplar is a creamy colored wood with swirls of green and sometimes a streak of purple in the grain like a final strip of a winter-sky sunset. It's an inexpensive hardwood and resistant to the dings and dents of a high-use space like a pantry. Hard- and softwood qualification has to do with how the tree handles reproduction. To raise ghosts from freshman-year biology: angiosperms, the ones that produce seeds with a covering, typically deciduous trees (the ones that lose their leaves), are hardwood trees. Mahogany, walnut, oak, teak, and ash are examples of hardwood. Pine, spruce, cedar, and redwood, coniferous trees, are softwoods, gymnosperms all. Their seeds fly naked in the wind. Softwoods grow fast, and are usually cheaper than hard. Hardwoods are typically denser (balsa wood, of those swooping two-piece airplanes from summer backyards, is an exception).

I measured, marked, and cut six shelves for each case and attached the trim to the outer edge. That made four boxes, two bases, twenty-four shelves, fifty-eight pieces of trim. A hundred and ten pieces of wood in total for these cabinets. Then came sanding, priming, and painting. From sheets of plywood and planks of poplar came four cabinets, solid things, useful.

"Hey Mary," I yelled from the porch. "Check it out." I stood there beaming next to the cases. Mary came out and smiled and gave me a high five. We didn't often touch or hug, and our high fives were awkward and sincere. I blushed. The feeling was genuine and unfamiliar—or not entirely unfamiliar, but coaxed from long-gone kid-like pride.

It was something more than that, too. Not just a look-what-I-did glee, but a truer satisfaction. By the end of the workday, I'd built four big cabinets, sturdy and square. Mary and I stood there together, both of us sweating. The sun sat heavy in the west, seeming to swell before it went about setting in earnest, and there was the feeling that something had happened that was right. First there was nothing, then there were cabinets. And these shelves would be used—for boxes of cereal and cans of beans, for cake tins and paper towels, for oatmeal, molasses, jars of spices. Mary smiled when we took a break on the back deck in the thick heat of late afternoon, and I told her that I loved those boxes. She laughed. "They look like double-wide coffins," she said.

The heat wave reached its peak a few days later, and the plumbers were swearing. The older one, Ben, with huge shoulders and a round, bald head, lay on the floor on his back, a thick forearm stretched beneath the kitchen sink. Sweat beads jeweled the skin of his skull. He closed his eyes as he felt for the pipes and the bolts, this large grown man on the floor with his eyes closed, sweat dripping off the smooth skin of his scalp. He closed his eyes to feel things better, and it made me think that maybe that's why we close our eyes when we kiss. When he lifted himself up, the dampness of his back darkened the slate tile, a shadow of sweat that dried quickly, like rocks on the beach in the sun.

The younger plumber, James, was in the basement shouting about water lines through a hissing walkie-talkie. Mary was in a crawl space above the kitchen. On her belly, she was working to align ducting that ran from the industrial-size oven vent over the stove up through the ceiling, across ten and a half feet of lightless crawl space, and out the exterior wall. When she flipped the switch to turn on the fan that would suck smoke and greasy fat bubbles up and away from the stove, it sounded like a jet taking off. Mary rustled above and dealt with the metal. When the plumbers weren't talking and the drills weren't screaming, and the hammerbangs halted, you could hear Mary humming.

The day began with talk of pigs.

"How're things on the farm?" Mary asked James.

"I've got a couple pigs now that weigh in over three hundred pounds. They're not much good for eating when they get much bigger than that." He talked of taking them to the slaughterhouse to get sausage back in one-pound bags. "You would not believe all the one-pound bags we've got. Freezers all over the place are filled with these one-pound bags." He doesn't name his animals, except his Saint Bernard. He had a cow named Meadow, and the Meadow burgers were delicious, "but it was a little sad," he said.

"Didn't you used to have some wild pig?"

"You mean that boar? Yeah, that mean thing." He had to bang it with a two-by-four once to keep it from attacking him. This was easy to picture: this big plumber with his bulging eyes and belly, whacking a wiry-haired wild-eyed beast with a club of wood. There's something wild-eyed about him, too, something impatient and sad. I liked hearing about his pigs.

"You still thinking about moving out to the country?" he asked Mary.

"I've been trying to persuade Emily we should buy a farm somewhere out Route 2."

"You should do it."

"That or Alaska."

We went about setting up the tools for the day, and the heat, even at nine a.m., felt like an opponent.

"There will be much swearing," Mary had said that morning as she opened the crawl-space hatch. It was one of her refrains. And it was an accurate forecast for that day.

I was in the back stairwell and my back was wet with sweat. Mary had given me a straightforward task: build a chase to cover up the pipes from the stair landing up to the ceiling.

"Chase?" I'd asked when she told me.

"Basically a column to hide the pipes. A tall, narrow, three-sided box." I looked at her blankly. "A pipe-hider." We stood on the back stairs and looked at the pipes, four of them, thick and thin, one of them cased in a foamy plastic cover. The guts of the house were peeking out, and a chase would make them disappear. "It's a chase when it runs vertical and a soffit when it hides pipes or ducts along ceilings," she explained. "First thing is to fire-stop the pipes," which meant spraying a toxic orange foam that swelled up like a burnt marshmallow around the holes in the ceiling and the floor where the pipes passed through. It came from a canister that looked like it might spray silly string. The foam hardens after it swells, and slows a fire's path as it burns through another story.

"No problem."

I made the measurements. I cut the wood. I fastened the three pieces together with glue and the nail gun. Mary was right: it was a long, thin, three-sided box. It was an easy day for me, especially compared to the hell Mary was in up above. I carried the pipe-hider over my right shoulder like an oar, careful not to knock it against doorframes or cabinets as I moved through the kitchen and down into the dark back stairwell.

I propped it against the wall. The entrance to the crawl space opened above me, and bits of insulation floated down as

Mary shifted. Some stuck to the skin of my arm. A fleck landed on my lip and I tried to spit it off.

"It's just newspaper," Mary said from above.

I didn't believe that. I imagined a mix of newspaper shreds, mouse piss, rodent-nest detritus, asbestos residue, and generalized cancer dust. I didn't want this toxic stuff on my lips, and my attempts to brush it off my damp forearms raised the worry that I was only mashing the poisons deeper into my pores. This was a regular sort of fear—when we mixed cement or sanded or stained and especially when we took down walls, I continued to fret about what was getting inside and the damage it would do.

"Can you bring me a flathead bit?" Mary called down. I was nurse to her doctor. I crawled up the ladder and shimmied through the hole. The heat of the space pressed in on me as if I'd been slotted into a toaster. Mary worked by the light of a camping lantern that she'd brought from her basement. Dust and insulation coated the skin on her arms and neck and face. I passed her the bit. She had a fearlessness when it came to her corporal self.

"I'm glad we waited for the hottest day of the year for this," she said.

"Sorry there's not room for both of us."

"No you're not."

"Do you want a mask?" I asked, knowing she'd refuse.

"I can barely breathe as it is."

I scrambled down and the duct metal twanged as she bent and attached one section to another. Ben the plumber struggled under the sink. He raised his work-booted foot off the floor to gain leverage. James was banging on pipes with a wrench. A clear ring of a bell clanged up from the basement.

I stood on the landing and raised the pipe-hider up and walked it back and forth toward the pipes. It covered them with just enough room on either side. I pressed it to get it flush to the wall, but it caught. A three-inch gap ran between the wall and the chase, from stair up to ceiling. Had I mismeasured? Had I gotten the distance wrong between platform and ceiling? I ran the tape up along the sides of the chase: a skosh less than a hundred and ten. I measured against the height of the ceiling. A hundred and ten on the nose. I leaned my weight into the chase. Nothing. It didn't give. I gave it a kick. It stuck firm.

Ben and James continued their walkie-talkie back and forth:

"You find it?"

"Found it."

"Over there by the furnace?"

"Yeah I found it."

"Everything okay?"

"Besides the fact that I'm sweating my dick off down here? Yeah, everything's okay."

Measure twice, cut once. The carpenter's proverb reminds us about planning, about accuracy, about the possibility of waste—of time, money, and material—when first steps are made with haste or distraction. "I cut it twice and it's still too short" was a joke Mary's old boss used to make, and I'd laughed when I heard it. Life is more forgiving than a two-by-four. Measure twice, measure six thousand times. I crouched and looked at the landing and I saw where my chase was catching. A swell at the seam of two floorboards—so slight—was bump enough to thwart the thing from fitting. Despite kicks, full body heaves, and all-my-weight pushing, the chase wouldn't move over the bump and press flush to the wall. By then I'd learned that measurement wasn't always absolute, that some-times a quick bash rightly placed could nullify parts of inches. The numbers say one thing, the flex and movement of wood another. Some pieces and places offered forgiveness.

Not here. Sweat dripped from my chin. The base of the chase needed shaving, which meant tugging it out, hoisting it back up on my shoulder, and maneuvering it back out to the deck where the tools were set up.

Bang. Slam. A knock against the doorframe.

"Use the sander," I heard from above.

The back deck looked out over the backs of houses in the Central Square neighborhood, which had its share of ne'er-do-wells, junkie congregations, piss smells on church doorways. It maintains a distinctly urban feel, a bit more grit and unpredict-ability compared to the rest of Cambridge, with its yoga studios and yogurt shops. The view showed small back gardens with swaying day lilies and bursts of hydrangeas. The old man next door spent each morning on his deck with the newspaper and a towering glass of orange juice. I waved. He raised his glass my way. He wore shorts and no shirt and the white hair on his chest stood out against his dark skin. A group of kids lived on the third floor across the way. Bikes leaned against their deck railing. They'd strung colored lights along the ceiling and used a milk carton as an ashtray. A girl in a tank top had a cigarette there in the afternoons. When we packed up around five those evenings, there'd be a few of them out there, and the hissing pop of bottle caps off beers made me thirsty for one too. An orange cat stalked around the patios below.

I sanded the corner of the chase where it hit the bulge, and the rest as well, grinding the wood away, wary of removing too much. Standing in a tub together tiling at some point early in our time together, Mary had said something that stuck in my head. "I look at wood the way I look at meat. You can always cut more off a piece of wood and you can always cook meat a little longer. Start rare with meat. Start long with wood."

Whenever I approached the saw, *start rare with meat* bounced in my brain, a measurement mantra. I cut the sander's power once a light flurry of sawdust coated the back-deck planks around my feet. I smoothed my fingers along the edges, surprised and pleased at how much like velvet a just-sanded piece of wood can feel. It tempted me to bend and rub my cheek against it, the same way, as a girl, I used to walk up to women in fur coats and rub my face against them. That splintery wood can be made to feel like velvet is a transformation I can't imagine growing weary of.

I hefted the thing back up on my shoulder and navigated back through the kitchen.

"Having fun?" Ben the plumber teased.

I positioned the chase back in place. I slid it toward the wall. It jammed again. I stood staring at it in silent, angry denial. The scream of a saw cutting through metal screeched up from the basement, then stopped. Ben's walkie-talkie hissed in the kitchen. What would happen if I just left, I wondered. What would Mary do if she came down from the crawl space and I was gone?

I tapped at the chase with my foot. I bent at the waist, one hand at the base, the other reaching up, like some awkward defensive football move, and I dug my feet in. With every muscle and all my will, I tried to push the chase over that little goddamn bump in the floor. Nothing. "Fuck!"

Mary rustled above. "Back it out," she called down.

I stood with my hands on my hips. I thought: this should be simpler. This is a simple thing, hiding these pipes with three pieces of wood. What Ben and James were up to, getting gas and water to flow through the house in the right way, or what Mary was doing up on her stomach in dim light, those tasks were challenging and important. My chase was cosmetic and it was chewing up the day.

Mary emerged from the crawl space and followed me out to the deck where she brushed the dust and insulation off her clothes and wiped her face. She looked like she'd come from a coal mine: dark dust circled her eyes, collected around the corners of her mouth, darkened the creases in the skin on her neck. She bent at the waist and roughed her hands through her short wiry salt-and-pepper hair. The dust moved off her in clouds. "Hold your breath," I said. She stood and rolled a cigarette.

I pressed the sander against the base again and she watched while she smoked and I ground the left corner so it would slide over the bump and fit against the wall. I lifted it to my shoulder as I had before, and Mary followed, holding the back to keep it from hitting the walls.

She stood at the top of the stairs as I placed the chase down again, raised it up, and slid it toward the wall. And, finally, it slipped in flush against the wall, tight up to it, up and down. The pipes disappeared. They'd been noticeable before, their thickness and color, and they invited one to wonder what was traveling through them: water, gas, or shit. It's surprising how thoroughly this wooden column camouflaged them. The chase vanished against the wall and wouldn't be given a second thought as Alice and Bettina headed down to the basement to fetch sweaters or a tennis racquet.

"Nice," Mary said. "Third time's a charm."

I was relieved to be finished. "It should've been a lot easier."

"You remembered to fire-stop the pipes?"

I closed my eyes. The little bottle of foam with its thin straw nozzle perched on a stair two feet away from me, where it had been all morning. Blood rose to my cheeks and my heart made its presence known in my chest, pounding in a way that said, *Get out of here, run.* It was too hot and the wick of my patience had already been burned from both ends.

"Son of a bitch," I said to the floor.

I hadn't remembered. I hadn't remembered despite Mary saying that it was the first thing I should do. I'd ignored the suggestion for two stupid reasons. One, I was eager to get on to the wood, to the more interesting part. Two, I hadn't felt like carrying the ladder to the stairwell and reaching up and spraying foam way up where the pipes met the ceiling. I'd wanted to get the wood done first, then fire-stop before installing the chase. Because of my laziness and childishness, I now wanted to take the chase and send it flying down the stairs. I wanted to kick the wall and crack the wood. I wanted to ask if maybe it'd be okay if no fire-thwarting measures were taken, just right here. I felt like a fool and I wanted the day to be done.

Mary laughed. For the heat and the frustration, the crawl space and the stairwell, for being covered in dust and sweat, for not getting it right, but for not fucking it up all the way either.

"Some days," she said.

And she was right. I was hot and sweaty and mad, but at least I hadn't spent the day rolling around in mouse shit, sucking newspaper bits into my lungs in a hundred-and-fifteen-degree heat.

This was one of Mary's greatest attributes. Mistakes weren't cause for scolding. Instead she used the fuck-ups (which were legion) to teach lessons. I envied her patience and often wished I could summon her calm and perseverance in the face of a stripped screw or a section of baseboard that couldn't be coaxed off the wall. Her wick of patience, especially when faced with the specific challenges of non-cooperation by inanimate objects, was miles long, could burn hours without going out. Some days you get it wrong, was the way she saw it, and let's see if we can figure out how to get it right.

The plumbers laughed when we reappeared in the kitchen. Mary was filthy; I'd soaked through my T-shirt, and scowled.

"Great weather," Ben joked. The two men wore work pants and sturdy lace-up boots and leather belts and long-sleeve button-downs with the company logo embroidered on the breast pocket. Their shirts were damp, too. There were challenges at every stage, Ben explained. They'd be back tomorrow.

"How much?" Mary asked.

Ben frowned, shook his hand give-or-take. "Twenty-two."

I didn't know much about plumbing except that it was expensive. Twenty-two hundred bucks seemed like a real deal considering all the surgery they'd had to do with rerouting the pipes in this old three-family place. "Not bad," I said.

Mary looked at me and said quietly, "Thousand."

My face showed shock and Ben gave me a wink.

Mary changed the subject. She joked about the hell she'd been in above the kitchen, and the guys wondered why Alice couldn't provide us with a fan. I wiped sweat off my face and tried to brush the sawdust off my calves. It clung to my skin like sand.

James with the belly and the mischievous eyes gave me a quick slap on the shoulder.

"Sure beats a desk," he said.

It did. But that didn't mean there weren't plenty of other days that made me want to scream the way the saws do, each with its own specific pitch.

The miter saw, also known as the chop saw, is the highest pitched. It makes a frantic, panicked, piercing sound. As the spinning blade is lowered into the wood, its desperate wail hurts my ears, even with the squishy orange plugs that we squeeze the tips of and twist into our ears. Of the saws we use, the miter seems most dangerous. It might be because the plastic guard that covers the blade as it spins has broken off so there is no protection between soft flesh and mean, spinning blade. Perhaps it's because it's the one we use most, the one

I've gotten most comfortable with, and therefore the one I'm most likely to be careless with. A momentary lapse in concentration could mean a finger on the floor, blood soaking into the sawdust. I try to remember this whenever I put my hand on the handle to squeeze the trigger to bring the blade into its spin.

The table saw has a lower, steadier roar. It's a saw made for longer cuts, for shearing the width of something long, like an eight-by-four-foot sheet of plywood. And when a piece of wood is run across it, the buzz is like the white-noise summer hum of walking in a field with lots of bugs making their warm-weather drone, a bug song in the heat. It's less menacing, calmer. That it's braced to the floor, steady on four legs, makes it less threatening. But every time we set it up, every time we crank the wheel to raise the blade so it emerges above the table, it brings to mind a torture device, a prisoner strapped nearby, gagged and thrashing, witnessing the blade rise. Scarier still is the image that comes when I'm standing in front of the saw, pushing wood across the surface of the table, and I imagine the blade dislodging itself from the piece on which it spins, flying out of its slot in the table and spinning at me, slicing me through, guts and spine, out the back of me, then rolling like a wheel across the grass, severing worm heads from worm tails, the remains left squirming in the dirt.

We use the table saw to cope, a process of carving out a section of crown molding, for example, so that it will press tight against another piece where the two meet in a corner. Because the table-saw blade is round and because the piece of wood has been sliced at an angle, the front of the piece of molding stays the way it is, and the blade chews back behind it so that the carved-out part follows the curve of the existing piece and fits snug against it, the way one hand cups around another hand's fist. With small, gentle movements, I press the piece of wood back and forth against the blade, and the saw takes something once solid and whole and firm and turns it to something multiple, divided and light enough to ride the air. I go slow.

I don't always cope well. Sometimes the blade chews a jagged bit on the outer edge, corrupting the smoothness of the curve, a mistake you can see from the floor, to be filled in and concealed with wood putty. Sometimes it's all going well, the movements are slow and smooth and right, and vision is locked in on the wood disappearing against the blade, and every other thing disappears except the line of the curve and the spinning blade and the sawdust rising, and then it gets away from me, the way a word, repeated too many times, turns to nonsense, its meaning lost. The line blurs, too much gets taken off, there's a nick in the part that's supposed to be smooth. I give myself extra inches so there is room to slice off errors and start again. Start rare with meat.

The jigsaw, handheld, its small thin blade rising up and down, is made for curved cuts. It delivers a thuddier sound, like someone trying to talk while running. It reminds me of a sewing machine, the saw blade thumping up and down through the wood like a needle through fabric. The Sawzall is a little like holding a machine gun with a blade on the end. When the blade hits wood, if your hold isn't firm, it bucks and jolts, and it takes strength to hold and press hard and pull the trigger-throttle to get the blade going at full speed. I don't like the wild-horse feel of it, the risk of buck and kick. Mary is able to tame it. For the ducting of Alice's giant oven vent, Mary had me hold and brace a long stretch of metal ducting with my hands and knees while she sawed through with the Sawzall. My whole body vibrated. My fingers tingled afterwards. I felt it in my elbows, this strange current, an energy transferred from blade to metal to my muscles and bones, a little like getting gently electrocuted, that strange bad buzz and tingle that makes you swipe your hand away from the power source.

T he saws didn't pose the only risks on jobs. Mary had to do some repair to one of the kitchen windows before Alice's slab of marble countertop could be installed underneath it.

"Holy shit," I said when I came into the kitchen. There was Mary, the bottom half of her anyway. She was on her stomach on the makeshift countertop, her legs sticking in from the third-floor window toward the kitchen, with the front half of her pitched down out the window. She yanked at thick caulk and pried away a bit of trim from the exterior wall. She'd hooked her thigh against the wall and was angled headfirst toward the ground.

"Do you want me to do something here?"

She shimmied back inside and stood up on the countertop. "I'm going to need your help," she said. "I need you to come over here and grab onto my belt."

"Mary, Jesus."

"It's fine."

I'm not afraid of heights, but I walked over to that window—which was less a window, having had its panes removed, and more a gaping rectangular hole in the wall—and I looked down those three stories to the brick pathway and metal fence below, and my head swirled and my stomach tightened. It became clear how thin walls are, how little separates us from the world below.

"Grab hold," Mary said. "I need to lean out and deal with the top corner." She knocked on the wall with the pry bar to show me where she'd be working.

I stood on the floor, tried to dig my sneakers into the tile, and grabbed the back of Mary's belt with both hands.

"You ready?" she said.

"Ready."

"Hold tight."

"Jesus!"

The leather belt around her waist was black and held up khaki cargo pants with big pockets down the legs. One pocket had a pouch of tobacco and probably a red plastic lighter. Another pocket had a utility knife and a few loose screws. The outside of the belt had a smooth gloss; the underside, a felted softness that might absorb moisture. By that point in the July afternoon, the sun had moved across the sky, leaving the metal fence and brick pathway below in shadow. The buckle on the belt was silver, but I couldn't see it at the moment. I gripped the back of her belt with both hands as Mary, standing on the counter, leaned her body out the window. My own belt buckle dug into the soft flesh below my bellybutton as I braced myself against the cherry cabinets we'd installed the week before. I felt the weight of her against the leather in my hands, my arms outstretched like Superman in flight. Though being suspended in the air was the last thing I wanted to think about.

Two fists could fit in the space between Mary's pants and back, and all but her legs had disappeared out the window as she twisted and cranked on the exterior wall. As a piece of caulk gave way, her body jerked with the give, and made my heart plunge into my stomach. A burning began in my arms, the gradual ignition and spread of heat under the load of her body, like the blinking on and brightening of tiny bulbs beneath my skin. What if she falls out of her pants? What if the leather rips? What if my brain goes haywire and I just let go?

"You okay in there?"

"Yep," I managed. "Hurry up, please."

She's going to die because of me. I started to imagine what I'd say to Emily. "*I dropped your wife out the window. I'm so sorry.*"

I leaned my weight back, concentrated on slow breaths. I could feel, with every molecule of my body, what it would be like were Mary to slip from my grip. I'd go flying back, thump down on the floor, she'd drop, I'd hear the sick sound of flesh and bone on brick like a sack of butcher waste. I couldn't imagine Mary screaming. I think she'd fall in silence. I saw myself leaping up from the floor, leaning out the window, seeing Mary's smashed body below.

I heard Alice behind me and turned over my shoulder.

"What's going on in here?" she said.

"The window," I rasped. I could tell my face was red with effort.

"Do you want me to hold on to you?" She started toward me, hands out to grab my belt. But I shook my head. "This is the dumbest thing I've ever seen," Alice said, hands up, looking vaguely horrified, backing away. "I can't watch."

"Almost done," Mary called out, as if she were watering a plant or dusting a shelf.

"I don't want to drop you." I couldn't hold much longer.

Finally Mary climbed back in the window. "Em would kill me," she said, sitting down on the counter, adjusting the waist of her pants, smiling.

My hands were shaking and I curled and flexed my fingers, stiff from the grip.

"I used to be a lot braver about heights," Mary said. "When I was your age I would've been hanging out there without someone holding on."

Alice returned and stood in the doorway, hands on her hips. "So I've got a stuntwoman for a carpenter." Unlike Bettina, who softened her imposing presence, Alice fluffed her feathers like an angry owl. She puffed herself up and threw her shoulders back. Her direct gaze gave her a force of presence that belied her five-foot-two height. "Let me be clear: I really just want a new kitchen. I do not want bodies falling out of my windows."

Mary laughed.

"You think I'm joking."

"That's nothing," Mary said.

I shook my head at Alice to say it wasn't nothing, that she and I were on the same team on this one.

"You're lucky you have such a strong harness," Alice said. I put my arms up in a joking flex.

"I'd rather not do that again," I said to Mary.

"All right, all right. No more adventures out windows." She looked down out the window. "That's a fall you could survive."

"You'd need goddamn wings to survive that fall," Alice said.

I n the familiar myth, Daedalus had wings—he and his doomed son Icarus. Daedalus was an artisan and inventor who fashioned wings of feathers and wax. He warned young Icarus: don't fly too low or the water from the waves will weight the wings. Don't fly too high or the sun will melt the wax. Both errors meant a fall. The middle way, like Goldilocks's porridge, not too hot, not too cold, was the rightest path. Father and son leapt from a cliff and flew like gulls. Intoxicated by flight, Icarus flew up and up. As his father warned, the heat of the sun melted the wax, the feathers fell, and Icarus fell with them. He tumbled through the sky, splashed into the sea, and drowned.

The less familiar prequel to the flight of Icarus and Daedalus also involves a young man falling. Daedalus took on his young nephew, Perdix, as his apprentice, and knew straightaway the kid was a genius. Pliny the Elder, in his *Natural History*, credits Daedalus with inventing carpentry. But it was Perdix who invented the saw. As Ovid tells it, when the two were together at the beach one day, Perdix spotted a fish's spine, bleached and jagged, on the sand. The boy touched the bones, pricked the sharp pieces against his fingertips, and discarded it for the seagulls to pick at, for it to dissolve into sand itself. The natural world supplied him with a vision. When Perdix got back to the workshop that day, he translated the pattern of the spine to an iron blade. He notched teeth into it, sharper than the fish's bone, and stronger, and so we have the saw, an essential tool for the very craft Daedalus helped create.

Daedalus, jealous of the child's gifts, couldn't endure being outshone. *He* was meant to be the master, the mentor, not this boy. Driven by envy, Daedalus shoved his nephew off a high wall of the Acropolis. With no wax wings to keep him soaring (and no one there to hold his belt), Perdix fell.

But not to his death. The goddess Athena, who favored crafts and smarts, had also recognized Perdix's genius, and she caught the boy mid-fall and turned him into a partridge. That squat bird makes its nests in low brambles and stays low to the ground when it flies because, as Ovid writes, "That bird recalls its ancient fall, and so it shuns the high and always seeks the low."

B ack on solid ground, it took two days to hang the upper cabinets in Alice's new kitchen. They were handsome, custom-built maple cabs the color of straw, unadorned in the Shaker style, and modernized with sleek silver rods to open the doors and drawers. I thought installing cabinets would be simple—just screw the things to the wall so they line up. It is exactly that simple, but it is not that easy.

Mary's standards for cabinet installation were uncompromising. A pair of Jorgensen clamps, the level, the drill, and some shims were the tools of those cabinet days.

The shims were flimsy wedge-shaped scraps of wood, thin sticks nine inches long and an inch and a half wide. I could break them in my hands. But they are crucial for leveling cabinets against bowed walls or floors. Usually made of cedar, they looked a little like the pieces of wood that come with a fresh gallon of paint. You slot them underneath and behind the cabs to straighten and level.

I acted as holder and shifter as Mary made each cabinet level and flush with the one next to it.

"Up," Mary said.

I'd press the cab up.

"Skosh more."

Another quick press with my shoulder and hand.

"Tiny bit more."

And again. "Too much. Too much. Lower."

And so it went until we got it right and Mary clamped it tight with the Jorgensens, checked again with her level, but mostly used her fingertips, rubbing the seam between the two cabs to feel for any bit of ridge where they met. The look and feel of one solid thing instead of two was the goal. Once achieved, we screwed the cabinet into studs behind the wall, and the partner cab beside it.

I got impatient with the time-consuming meticulousness of it all, the fractional shifting. "It looks *fine*. No one is ever going to be able to notice that it's off."

"I notice," Mary said. "Someday you'll notice. Can't live with it. Sorry."

We were in the midst of it when Mary's phone rang in her pocket. I was glad to rest my arms.

"Has it really been four years?" she said into her phone. She shrugged in my direction suggesting she wasn't sure why this guy was calling. It got clear fast.

"What sort of health problems?" she asked. "Ah, Kev," she

said. Her face changed. Her voice got lower. "Ah, Kev, that's bad." She laughed a little bit, and answered questions about her daughter. "She's a teenager, can you believe it? She's starting to act like one, too. That's the scary part." More laughter. "That's one way to do it," she said. "That's definitely one way to do it."

After a bit more back and forth, she hung up the phone and went to the back porch for a smoke without a word. I stood and looked at the tiles, saw the sawdust I'd need to sweep from the grout lines. The slab of marble countertop looked cold to the touch, smooth white with lines of gray and black that flared and streaked like nerves. The sun glinted off the metal basin of the KitchenAid mixer that sat on the marble by the window.

"I think he was calling to say goodbye," Mary said when she came back to the kitchen. She explained he was an older guy, that they'd worked together on the same carpentry crew for years under her old boss, that he lived in Pittsburgh now. Cancer everywhere, tumors everywhere.

"Not good," she said.

Coffins cost four grand, he'd told her. But carpenters know other carpenters. He told her, "I got a buddy who's going to make me one out of plywood and two-by-fours. Two hundred max."

The weeks passed at Alice's and day by day the space became a room again. I'd noticed it at other jobs, the shorter ones we'd had, but it was especially pronounced here: whenever we worked on a place, it became ours. Once we got in there, got the tools set up, started whatever it was we'd been hired to do, it felt like we owned the place. In his story "The Swimmer," John Cheever described the ownership a lover feels over his mistress's property. "He stepped through the gate of the wall that surrounded her pool with nothing so considered as self-confidence. It seemed in a way to be his pool, as the lover, particularly the illicit lover, enjoys the possessions of his mistress with an authority unknown to holy matrimony."

Substitute a kitchen for a pool, home ownership for holy matrimony, and a relatively inexperienced carpenter for an illicit lover, and this is the sense of entitlement I felt in the places we worked. With nothing so considered as self-confidence, the kitchen became our kitchen, the hall became our hall, the deck became our deck. The owner, for the duration, would be in some way severed from the place.

At Alice's the saws lived on the back porch. The ladders leaned in the dining room. The hallway floors were papered and taped to keep the dust and grime off. And when Alice came into the kitchen to make a sandwich or steam up some dumplings, I'd think to myself: *Scram, lady. This is our place right now.*

We'd been there for a little over a month when Alice started to reclaim the space. The pipes had been shifted and the plumbers had moved on to another job. We'd tiled the floor with those lovely gray tiles from Italy. We'd rocked the appliances into place. We'd hung the huge vent over the stove so Alice could grill meat in the house. Mary had made sure the cabinets, upper and lower, were perfect on the wall and on the floor. The pantry was built, with sliding doors on the cabs.

We came back one morning to find that Alice had put

cans on the shelves in the pantry. The next day, cookbooks filled the shelves on the left wall. A drawer had silverware, another had spatulas, graters, and long silver skewers. A tea-kettle arrived on the stove, a basket of apples on the counter. The room was a functioning place again. We were almost finished.

One of the final steps involved hanging a monster of an exterior door that led from the porch to the pantry.

"This door is a beast," Mary warned. Hanging doors can be a real pain in the ass. Precision is required, or else the door won't swing on its hinges or click in its latch. It'll scuff the jamb, stick in the frame, require tugging to open and a hard slam to close.

Out on the deck, I crouched by the big green door's base. Mary was on the other side in the pantry. We couldn't see each other, could only hear each other's voices. The right corner on the hinged side had to be raised a bit on the sill of the frame. I was trying to lift and push. Mary was ready with shims on the other side, to slot underneath to keep it in place while it was screwed into the frame at the right height. I could not get the door to budge. The distance up I was aiming for: three-sixteenths of an inch. I strained and swore quietly and could get no leverage.

I pressed, face crimson with effort, and the door remained exactly where it was. *There's no way I'm going to lift this*, I thought.

I repositioned, inhaled, tried again. Nothing.

I lost patience.

In desperation and frustration, I opted for brute force. I strained with all I had, every muscle engaged, stoked with frustration, enhanced by anger. A surge of effort and strength. And then—the door lifted! Lurched! A miracle! I'd done it!

And from the other side of the door I heard, "Fuck, fuck, fuck."

In going too fast and too hard, I had badly jammed Mary's thumb.

I'd seen her bang herself dozens of times. She usually cursed and made light. News that she was bleeding would be relayed with the same nonchalance as she'd announce that we needed more galvanized nails, a simple transmission of fact. I had asked her one morning if she'd ever cried on the job. She'd given me a look that made me wish I hadn't asked before saying, "No."

She wasn't crying now, but cursing, hard. "Fuck," she said again. Then she got quiet and began to roll a cigarette.

I apologized and put my hands over my face. She looked at her thumb, wiggled it, blew on it.

"You got me that time." I apologized again. "Probably going to lose it," she said of her thumbnail. She shrugged. "It'll grow back."

When she recovered and we'd gone back to wrestling with the door, she said from the other side, "Okay, all right, *finesse*, girlfriend. Lift a little, *slowly*."

I lifted, she pulled, and we got it. The door moved right on its hinges, swept across the threshold, closed with a satisfying *swump-puh*.

Two days later, Mary handed me a hammer and her thumb was pure black, as though someone had injected ink underneath her nail.

"Oh Jesus, Mary. I'm so sorry."

"It's pretty tender," she said. "But I don't think it's going to fall off."

It took me a long time to learn that not every problem on the job—in fact, very few of them—could be solved with brute strength. When a piece of trim wouldn't come out of the window frame, or a cabinet wouldn't settle to level, or when things seemed jammed or stuck or too thickly glued or caulked or impossible to square or non-cooperative in whatever way, it was my impulse to opt for muscle over mind, to accomplish what needed accomplishing by force of body and will. My reservoir of patience was shallow and quickly drained. Thwarted, my mind got tight and hot, and body reacted in kind, fast and dumb.

In this way, I broke things. I broke drill bits, pieces of trim, glass. I put dents and punctures in walls and floors, and, in one unfortunate slip, the soft meat of my palm where a scar marks the spot.

"Finesse," Mary said often. In other words, be gentle, go slow, don't rip and yank with all your might. Let the materials tell you which way they want to go. Use your mind and listen closely. Allow physics and the tools and patience to get the job done.

"It's all about coaxing," Mary said. "Knowing where to put the pressure."

The final days of a job involve lugging and loading. Striking items off a final punch list, cleaning before we leave it for good for the owners to pile their lives in their new, changed space. We would walk through a place, once, twice, three times, to make sure the dust was out from between the tiles, the bit of paint around the light switch got touched up, the small hole in the stair was patched and painted black like the rest. On the last day at Alice's, we reviewed the punch list, nailed in a few last pieces of trim, slotted the huge chest freezer into the bay we'd built for it in the pantry. We swept, dusted, vacuumed, mopped. Done, and done. All day I couldn't help but talk about how great everything looked. I went around knocking on the doors of the cabs. I rubbed the smooth white knob on a door in the pantry. I pulled out drawers ("Stop snooping," said Mary). I felt excited and I felt proud. Mary felt it, too. She didn't go around yammering about how amazing it all was, but she was smiling. We stood in the corner of the kitchen, the same place we'd stood on the first day there when the room was blank, and took a good look. Mary patted my shoulder. "Nice job," she said.

So much work went into this kitchen. So many hours and so much sweat and effort, and we'd never see it again. It was ours—and not ours at all. All that was left to do was hang a screen door from the pantry out to the back porch. It was light and flimsy like a summer dress. We rigged it so it closed with a hush, a quiet tap against the frame. Perfect.

Alice appeared behind us. "No, no," she said. "I want it to slam. It reminds me of summer."

So we adjusted the spring and Mary and I stood in the pantry and we watched the door swing closed and we both blinked when it slammed like summer.

Chapter 6

———

LEVEL

On shifting, settling, and shifting again

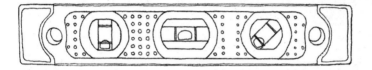

After Alice's kitchen, we spent five weeks on the third floor of an old Victorian place with a turret where we moved walls and put down a hickory floor. The hickory, such hard wood, put up a good fight against the miter-saw blade, and gave off a rancid smell when cut, not sweet hickory-smoked barbecue, but something bile-like with a sour sting, likely from whatever chemical finish was used on the wood (*Contains chemicals known to the state of California to cause cancer* it read on every box of boards). We did a kitchen at a place outside Harvard Square in Cambridge for a couple who became my favorite of all the people we worked for. They left us lemon cookies on the counter with notes that said "Eat these," gave us jars of homemade jam from the raspberries and blackberries from their place in Vermont, joined us when we broke for lunch. The transplanted jade plant the woman gave me grows by my living room window. It wasn't just their generosity: I liked the love the pair had. In their late fifties, early sixties, there was an amused exasperation between them, a sense of patience, an evident, affectionate closeness. "We want you to adopt us," Mary told them.

We did a quick deck in Arlington, a historic suburb that edges up against Cambridge, on the route of Paul Revere's ride. It took four days.

Somewhere along the way, I became the communicator in our team of two. Mary, besides talk of fleeing human company for Alaska, spoke fluent carpentry but forgot at times that not everyone understood her language. In situations where she outlined some aspect of the work to the client and got slow nods of semi-comprehension in response, I translated, with my beginner's understanding of the grammar and vocabulary.

This: "We're going to sister up the stud and patch the wall. The mud'll dry overnight and tomorrow we'll frame the cabs."

Became this: "So we're going to attach another two-by-four to this one here to give it extra support, then close up the hole here in the wall. Mud is drywall compound?" I'd say, to suggest, *Maybe you know this? Or maybe this is a weird quirk of Mary's to call this mud?* ". . . And it can take about a day to dry. And we'll be building the boxes for the bookcases, the outside shell of them"—using my hands to outline a rectangle—"tomorrow."

Mary sometimes thanked me.

"I just think of what I wouldn't have understood," I'd say, which was everything. And I'm sure I've offended a homeowner in offering up an explanation of something that he or she already had full understanding of. *I know what* joists *are.*

Over these weeks, which piled up quickly to months, my muscles came back. I'd flex in the mirror, glad to see definition again in my arms. And Mary and I found our rhythm again.

A summer spent in kitchens and turreted third floors, on front porches and in closets lined with cedar, soon gave way to fall. Mary asked if I'd be willing to help on a project in her own house.

Her place was a work in progress, ever unfinished. The wallpaper up the narrow, twisting back stairs was faded and tearing. Holes in the plaster marked the walls like picked scabs. Plaster crumbs crunched under foot. Trim had been pulled from around the doors in a bathroom; the tub was a mess of old paintbrushes and kitty litter; the floor was battered oak; bits of prickly lath poked through holes in the bathroom wall. In the living room, where the chimney had been removed a couple years before by the wild demo man and his two sons, an ill-fitting piece of plywood covered the hole in the floor. Blue painters' tape remained stuck to the ceiling from when we'd taped up plastic sheeting to keep brick and soot dust from drifting into the kitchen when the chimney came down.

The house was full of jobs half-done, and it stood in contrast to the meticulous way Mary tended to her worksites. Her own house was unfinished because paying work called again, and time rushes along. At first the kitchen had trim on only one wall for a few weeks, then a couple months, and then a year had swept by and there was blankness where the trim should be and maybe it had been forgotten, had disappeared into the familiar landscape of the house. Or maybe, with every time she entered, with every soup simmered, with every dish washed, a glance was given to where the trim should be, another reminder, *that, too.*

Mary was committed to a room on the third floor, a space of peaked ceilings, old dark wood, and small dormer windows. She wanted to make that room her office. It was dim, dusty, and cramped, with wood scraps and a big trash bin and rolls of insulation littering the floor. The two of us bumped into each other as we negotiated the small space, made smaller by the saws and tool buckets.

The ceiling slanted with the sharp rise of the roof. The house was built in 1886, and the boards were broad and dark. We framed a knee wall, about three feet high, used to support the rafters and named after its approximate height. We ran wires for lights at the top of the room. We shoved candy-pink insulation into the bays between the joists.

"Not there," Mary said, as I reached to press another strip into a dark bay. "We're going to do a skylight."

We cut away the wood that ran across the rafters with the Sawzall—splintery old one-by-three—and pulled out old nails. We made the rough frame, cut two-by-tens and ripped them to match the wide rafters. Mary sliced through wood and roof to create the opening. A small flap at first, and there was sky, and light fell onto the floor as though it had been shoved, and cold air spilled in like water. Mary climbed out onto the thin lip of roof and crawled up and cut more of the opening from the outside, crumbs of roof tile rolling into the gutter. More light hit the floor. The room changed. And the same sense I had at the Russians' house, where bugs had feasted on the wood of their bay window frame, returned: what if we can't close it up in time? The days were shorter then. What if it rains? But those questions were quieted. We've done this before, a piece at a time. We'll do it again now.

Mary went about removing the roofing tiles, prying them up without tearing them, pulling up the wide-headed roofing nails. I lifted the window itself out through the hole to Mary's hands as she knelt on the roof. We placed it, made sure it was centered, shifted it back and forth within the hole, up and down, and Mary hammered in a few nails to keep it in place. It was a cold November day, bright and cloudless, and Mary puffed on her hands and clapped to keep them warm inside her work gloves. She installed the flashing and went about the slow process of re-installing the sheets of roof.

"I'd never want to be a roofer," she called out.

While she was out there on the roof in the cold, I nailed lath to an arched space that led to the other window in the room. She hammered down on the roof. I hammered up into the ceiling. We pounded and pounded. Down and up, putting the room together, changing it and keeping it the same. She climbed back in through the window, pale with cold.

"All finished out there?"

"All finished."

The sun, which had started at Mary's back as she worked, beaming into the room, had shifted across the sky, was behind the other side of the house now, and setting. The sky softened to a purple and a few thin twists of cloud like cigarettes side by side rolled high above the horizon.

"It's off by half an inch," Mary said. I was silent and wondered if this meant taking it out, doing it all again. It must have slipped a bit before she nailed it in, she explained. "I should've thrown the level on it. You want to go see if we can tell from below?" I dreaded the look.

"Will it leak?"

"God, no," Mary said. "I fucking hope not. I'm just worried it's going to look crooked."

She stopped by the woodstove to warm her hands on our way to the porch on the second floor. We looked up at the roof, at the new window. And what a relief—no evidence of the half-inch offness. It looked as it should.

Back upstairs, we stood side by side in front of the new window, shoulder to shoulder, and looked out. To the back porches and peaked windows of the houses nearby, lights just beginning to be flicked on, that inviting soft orange-golden glow against the dimming sky. To the Boston skyline three and a half miles away, a purple blur of buildings, rise and fall, above the stretch of Charles River we couldn't see, and the bridges that spanned it. To new clouds, loose puffs, steady and high.

Red leaves clung to the tall maple out the window, leaves that would drop in the coming days, not long for this world, leaving the skeleton of the tree against the sky. I've always loved November, when the bones start to show.

I did my usual gushing—*it's amazing, look, incredible*—astonished again at the work and its power to transform. The room remained unfinished, still a mess of guts and wires and wood, everything exposed, dust everywhere, thistles of insulation in the air. But it would all get done. In weeks or months. Slowly it comes together. The room was different now. It went from dim and cramped to bright-lit and inviting, a good place to sit and get some thinking done as the sun describes its arc across the sky, as the leaves fall off again to end another season and the tree adds another ring.

Mary nodded. "This changes everything."

The maple leaves dropped, the temperature fell, and we slipped into winter. After the skylight, in the slowing of the year, Mary planned to pause the progress on her third-floor office space in favor of redoing a bathroom downstairs, the one with the paintbrushes in the tub and the crumbling walls.

I swung by her place to pick up the last check she owed me before we took our annual break. She walked me through her bathroom plan.

"Give me a call if you want some help," I said.

"We'll see if I can afford you. I'm scared shitless about how much the plumbing is going to cost."

We parted ways with a hug and Christmas wishes, knowing it might be some months before we paired up again. I didn't fear the slowing. I knew next season would come.

Around this time, my father and his girlfriend bought a house together in the woods by a tidal river in southeastern Massachusetts. My father finally collected his belongings out of the storage bay they'd been occupying for six years. To visit his new home was to see the familiar items from our growing up freed from dark boxes in a storage cell. Many boxes stacked in the basement remained to be unpacked, most of them labeled BOOKS.

During one of the first visits there, we sat near the fireplace, my brothers, father, and I, and our respective romantic partners. Outside the window, the bird feeder was a flurry of action. Tubby morning doves, bright darting cardinals, feathers a duller red than their full-force summer color, a nuthatch, some chickadees, a woodpecker. They fluttered and fed, some pecking at the feeder that sat atop a pole, some on the ground picking at seeds, some at the small cage of white suet that hung from a branch, cow fat white like snow. My father identified each bird. When one swooped in to scope the scene from the branches nearby, he would forecast which feeder the little bird would go to—pole, ground, or suet. He was right every time. He talked about how you could feel the presence of a hawk nearby—the birds would still, then scatter.

After watching the birds, we turned our attention back inside, toward the fire. Darkness settled, the window out to the feeder reflected the lamps, the stone fireplace, our faces. We chatted and laughed. Finally everyone started toward bed. My father stood, looked at me, and raised his hands toward either side of the fireplace.

"Bookcases," he said toward the enormous blank spaces on the wall. I could picture it immediately.

"Great idea," I said.

"I'd like for you to build them."

I frowned. The relaxed feeling brought on by an evening of fireside laughs shifted to a storm of doubt. For me to build them? By myself? I could not say out loud that I wasn't sure I could, that after these years with Mary, I doubted my ability to build cases on my own. I did not want to admit that the thought of it scared me. So I lied. I told him I wasn't sure what my schedule was with Mary these days. "I don't know if I'll have the time."

I went to bed that night and thought about the cases. My reaction when he'd asked was immediate and surprising. Could I? I knew how to do this, didn't I? I went through the steps in my head, the ones I'd learned from Mary and done with her many times. I pieced the cases together mentally, starting with the bases on which they'd sit, moving on to the frames, the shelves, the trim. I'd have them match the height of the window trim, I thought, keep that line consistent around the room. An outlet on one wall would mean notching a hole in the back. These were all things I'd done before, had seen Mary do.

"Keep me posted on the shelves," my dad said as we left. "I'd like to start unpacking those books."

Back in Cambridge, I kept thinking about the cases. I made them more real in my head, more possible. The floors probably aren't level, I figured, and reminded myself how to correct for that. I'll have to alter the trim around the window. In my mind, I knew what to do.

But when my father called to see if I was up for the project, again I hesitated. The project had been taking shape in my thoughts, but, tools in hand, would I be able to translate what I knew to the wood? Without Mary at the helm, would I come to discover that I'd learned nothing? A terrible thought, it brought a clenching sort of discomfort, the confrontation that I'd been living a lie. I could dress the part, but did it mean I could do the work?

The feeling was familiar. When I began my job at the newspaper, when I first started filing stories, I'd wake up before work in anguish. How am I going to do this? What if I don't finish on time? What if I can't figure out how to say what I want to say? It was a specific, potent fear of failure, of being struck with the inability to express what I knew, or to do so in a way that revealed me for the faker I was.

The carpentry questions echoed the journalism ones. *How am I going to do this? What if they don't work? What if I can't figure out how to make them stand? What if I can't translate what I know to the wood?* Doubt crowded my thoughts and delayed any possible start. To begin was to open the possibility of fucking it up.

The novelist Gabriel García Márquez once told the *Paris Review* that "ultimately, literature is nothing but carpentry. . . . Both are very hard work. . . . With both you are working with reality, a material just as hard as wood."

It's true that writing and carpentry both require patience and practice, and both revolve around the effort of making something right and good. Both involve getting it wrong over and over, and being able to stay with it until it is right. In both, the best way of understanding something is often by taking it apart. In both, small individual pieces combine and connect to make something larger, total, whole. In both, we start with nothing and end with something.

But what appealed to me so much about carpentry work is how far it is from words. The zone of my brain that gets activated building bookshelves is a different one than the one that puts together sentences. And what a relief it can be, not having to worry about the right word, not having to think, over and over, is this the best way to say this? The questions carpentry raises are the same, ultimately—will this work? Will this function as it should, be true and strong? But the answers come from different rooms in my head, and it is good to exit the word room in favor of a less-used realm that deals with space, numbers, tools, and materials. Much of what carpentry requires does not come naturally to me. Angles, numbers, basic logic. But with carpentry you have a tape measure, a saw, a pencil, a piece of wood. Concrete, understandable, real in the world, each of these things made for a specific purpose.

García Márquez admits a few sentences later that he'd never done any carpentry himself. If he had, he'd know that a piece of wood is not the same as words. A wall is real. A piece of baseboard that hides the gap between the wall and the floor, that's real, too. There's a sense of completion with carpentry that doesn't exist with writing. Words are ghosty and mutable. A measurement, a cut, sawdust in my lungs, and the piece of wood slides in to fit tight with a few taps of the hammer. It's the opposite of abstract. Measure, measure, mark. Cut. Nail in.

The process of building a writing office in his Connecticut backyard reminded Michael Pollan "just how much of reality slips through the net of our words." Language becomes less useful when you're building a bookcase. A certain head-emptying, in the finest moments, takes place. That meditative goal, rising above the words and emotional swamps, being fully awake to the tools and the wood, involves the evacuation of language. What a relief it can be, for words not to matter. The shelf is real, and right now, as I sand it smooth, it's all there is. To write is to muck around in the space inside your skull. It is to build something, yes—worlds and people, moods and truths—but it is closer to a conjuring. You cannot put your wineglass down on a paragraph, even if that paragraph is perfect.

Much of what Mary taught me did not involve words. The classic writing dictum applies in carpentry, too: *show, don't tell*. It's hard to explain how to install crown molding. It's best learned by watching it done, and doing it. Over and over. Her verbal lessons—*start rare with meat; finesse; go slow; be smarter than the tools*—are all enacted in the way she does her work, the way she moves and uses her tools to solve each problem. You could read books and books on how to build a wall or tile a floor, hear someone speak for hours on the best ways to make a bureau or a bookcase. They could use all the right words, weave the tightest net, but until you grip the hammer in your palm, until you feel two pieces of wood pressed flush against each other before they are fastened, until you stand back from what you've made and then walk up to it and kick it or place something on it, you will not know how it's done. All the language in the world won't make a bookshelf exist. It takes watching, and doing, and screwing it up, and doing it again and again until it is done.

I spent hours sketching and adding and subtracting in planning out the bookcases my father wanted. I called Mary to see if she could loan me a few of her tools.

"You're heading out on your own," I could hear the smile in her voice. "Good for you."

"I haven't said yes yet."

"Say yes! You know what you're doing. Remember that it's going to take longer than you think."

"I was guessing four days?"

"I'm guessing more like eight."

"Shit."

"Remember when you could barely use a drill?"

That night I dreamed about bookcases. Up on a ladder in the sand, I was building a bookcase on a beach. The shelves faced the sea and the tide was rising, waves washing in to lower shelves, soaking the books that were already filling them, making the pages swell, drawing some off the shelves and back into the ocean. I was building the bookcases higher and higher so they'd rise above the biggest waves. When I turned, I saw seagulls dive-bombing the books that had been swept away and floated on the sea. My ladder kept slipping in the sand. Dread: how will I hammer through water?

In the morning, I called my dad and told him I was available for the project.

It was the third day into the first real cold snap of the season, and the dry tight cold made everything seem brittle, bones and branches. The highway on the drive down, with the wood loaded into the car, seemed bleached by the cold. The sky was pale.

I arrived late in the afternoon, pulled down the dirt driveway with spindly trees closing in on either side, tall and narrow-trunked, fuzzed with pale green lichen. The house had the feel of a cabin—woodstoves and wool blankets, a high-peaked roof. The air down there had the sweet mulchy stink of wood and dead leaves, a whisper of the sea. Coming from the city, I noticed the quiet. A chatter of birds, rustle of branches and dried dead leaves. There was no city hum, no low rumble and buzz of traffic, movement, streetlights, no static of a neighbor's television. Here, at night, the darkness and silence collected around the house like a quilt.

I unloaded the wood on the thin rim of back porch that faced three feet of grass and a wall of mossy forest and the river somewhere beyond. I looked at the stack of boards, the bundles of trim, and it seemed impossible that these would come together to make something real and useful. The light was fading and I stared at the wood, imagined the way each board and stick of trim would be cut, how they'd be fastened. Great steamy puffs of breath rose around my face with every exhale.

In what little light remained of the day, I drilled the holes in the case sides where pegs would slot in to hold up the shelves. I held the drill and stared at the wood some more. I took a deep breath, knowing that this first hole was the first chance to make a mistake. To look is to keep it perfect in your mind. To take the tool to the wood is to open yourself up to error. *You know how to do this*, I told myself. I placed the drill and squeezed the trigger and the bit burrowed down into the wood. The noise against the quiet of the marsh almost seemed a violence. I drilled hole after hole. Ducks made noise on the river. I'd finished the holes on two boards, half done, when it started to snow. The porch lamps bathed the wood in light. And if I held my breath over the planks of wood, I could hear the sound of snow falling, a papery whisper.

My hands ached from cold by the time I finished with the peg holes. I stacked the boards up, set the sawhorses aside, and put the drill back into its case. I shivered a bit inside. I'd be staying here until the shelves were done. Besides plotting out the shelves over and over in my mind, my thoughts kept returning to my father's inevitable criticisms. He is a perfectionist, and quick to call out fault. I imagined him hovering over the work, in his khaki pants and leather shoes and layered shirts, clicking his tongue. *You're doing it like that?* I anticipated having to remind him that he'd hired me.

I stomped around the living room to shake off the cold and he came in and told me to sit down. The gravity of his tone caused the rise of walls, the ones that shoot up to protect against coming bad news, to guard against the things you don't want to hear. I sat and looked at my lap, pretended to focus on thawing my fingers.

"You are the boss," he said. "And you have the right to kick me in the shins if I start being an asshole."

I laughed. This was not what I'd expected.

He said it made him happy to have me be the one building these things, putting my stamp on this new house in this new phase. He talked about pride. He talked about how much it meant to him. How to explain the discomfort provoked by this moment of sincerity? This was not how we communicated in our family. We made jokes and talked books, and affection was understood as opposed to expressed. As he spoke, I tried telepathy: *Stop, please, even this is too much.* I glanced at him. *Oh no please are those tears in his eyes?* I felt shy and eager to rush away. So I scoffed and dismissed it with a shrug. "We'll see how they turn out," I said from behind the walls. But from there, I felt the significance of these shelves, too, of contributing to this

new home and phase by making a home for his books. He loved
to quote the Anthony Powell title: "Books do furnish a room."

We had soup for dinner that night, thick soup he'd made
with sausage and red pepper and white beans. We ate it sitting
side by side at the kitchen island. It was exactly the sort of food
I wanted after standing in the cold. He warmed the bowls with
hot water before serving up the soup.

He saw me flipping through a stack of seed catalogues left
on the counter, something I remembered from childhood,
looking at all the colorful pictures of pansies and melons and
zucchinis, and all of them and more appearing in our backyard
in summer. "We're going to clear some trees on the south side
of the house and make a garden," he said.

As we ate, he talked of decoys. He talked of having a workshop again. He'd been unpacking his tools. The workbench in
the basement had a scatter of clamps and bullet levels, paintbrushes, half-carved shore birds, pale and paintless, pieces of
driftwood, files, chisels, and rasps, all those wooden-handled
tools I still didn't know the names of, all of them freed from
boxes finally and ready to be used again. I bet it felt good for
him to have his hands on these tools, to feel the wooden bodies
of the birds, to feel the potential, to start to carve again.

"Stay here," he said, after we finished our soup. He went
down to the basement and I heard rustling from below. "It's
amazing the stuff I'm coming across," he said on his way back
up the stairs. He returned to the kitchen with a cardboard tube
under his arm and removed a scroll of crinkly delicate tracing
paper, dry and faded a tea-stained yellow. He unrolled it to
show a pencil drawing of a great blue heron with its S-shaped
neck and stalky legs, a beautiful line drawing life-size at nearly

four feet high. I'd thought he'd long forgotten his promise to make me a heron out of wood. "Pretty cool, isn't it?" my dad said. I told him it was extremely cool. "Now I just need to translate the drawing into wood. Imagine it in three dimensions."

I t didn't warm up any overnight. In the morning, I made the boxes, the outer shell of the cases, and fastened on the backs. I cut the shelves, six for each case, and cut the pieces of trim to line the shelves and the cases, too. Measure, mark, cut—again and again. I attached the pieces of trim to the shelves, made the strips of poplar one-by-two flush with the top of each shelf to hide the unfinished edge of the plywood behind it.

My father went about his day, drinking big mugs of tea and working at the computer on a marketing strategy for a Boston nonprofit. And he watched his birds at the feeder outside. "There's a woodpecker," he'd call from the other room, "another downy," and I'd lean to look out the window and see its red head and black-and-white-flecked wings. Its cheerful tap of beak against wood drummed out through the forest.

I moved on to sanding, priming, painting, which seemed to last for days. My boyfriend Jonah joined me for the last stages of the project, and it was good to have the help and company, to break the tedium and speed the sanding, priming, painting process. I had nerves for the eventual installation, when errors would reveal themselves. I had sent Mary a few panicked texts. *What happens when —? Do we do it this way or —?* And she wrote back straightaway with simple answers.

The floors bowed, rising and falling like low-tide waves.

They required time with shims and the level in order to right the bases on which the cases would sit, raising and lowering them until the bubble in the level slipped between its lines.

The levels with tubes almost full of yellow or chemical green liquid and an air bubble that slips back and forth inside are known as spirit levels. The alcohol in the tubes gives the spirit level its name. The laying of the level is one of the final tests of a carpenter's work—the bubble settles itself in the middle. Perfect, yes: clamp, screw, check again; still level? Good, done. Press it against a doorframe, up and down, and the bubble finds center if all is as it should be.

I sometimes wish a tool existed that could measure the plumbness of our spirits, a tool that would help us decide what's right for our own lives. How helpful to have an instrument that signaled, with the silent fluid shift of a bubble, that we should shift our spirit a little to the left—just a skosh— and all would be balanced and right. It's not like that in life, of course. If your spirit is level one minute, there's no guarantee it will be level the next. We shift, or don't, make adjustments, change, with the intention and the hope—and sometimes nothing so intentional—that the bubble will find center.

Mary had a six-foot level, but we mostly used the two-footer and the bullet level, a little guy, six inches long. The levels have three tubes, one in the center and one at each end. The center tube reads for level on the horizontal: a floor, a shelf. The ones at either end measure for plumb on vertical readings, a doorframe, a wall. Two tiny lines mark each tube, and the bubble inside is exactly the size of the distance between those two marks.

It's a silent tool. To see that bubble land between the lines is to feel relief and satisfaction. It's a tool that's also brought about temporary lapses in sanity. In adjusting cabinets on the floor, for example, a thin shim in the front corner gets the side-to-side reading right, but throws off the front-to-back. More shims, more adjustments, space fragments up and down. I lose the way. It's a similar feeling of being so close to a piece of writing that suddenly you can't see it, the plot goes, the whole thing vanishes, there but unseeable. The same happens some-times with leveling. The bubble shifts and settles but refuses to tell you what you want it to tell you. A shim in and out, another, and nothing's where it should be and each move gets you further away from where you want to be. I've had to step away, to approach another task, empty my head, then come back to leveling, removing all my little stacks of shims and starting fresh, to try again from scratch.

Once the bases were level, it was time to put the boxes up against the wall to see if the fit was right. I feared this moment. I feared the miscalculations that would be revealed. The first one, to the right of the fireplace, fit just right. It was the simpler one, the one that didn't edge up against a window. I was pleased with the distance between the light switch and the side of the case, and pleased that it fit as it should against the stones of the fire-place, too. I pressed the other one into place. The outlet hole I'd made with the jigsaw slipped over the outlet right on center. The left side was flush against the piece of window trim I'd had to remove and rip to make the case fit. Oh, the seam was perfect! I marveled. This is always the moment—before it's all finished, before the last piece has gone in and you're tidying and on your way out—when you can really see it, when it feels the best.

My father took a break from his work and came into the living room as I stood back and looked at the cases, hands on my hips. His smile was big and genuine. "Hey, all right," he said. He gave me a high five. He could see it, too.

Later that afternoon, as I tidied up the tools and stowed the paint cans for the day, buzzing with relief that the cases fit, that I'd done it right and well, my father came into the room, dark news written on his face. He'd just gotten an e-mail from my younger brother, who'd written that his girlfriend's father was dying, and it was happening fast. In the years my brother and she had been together, she'd become a good friend. Her big laugh upped the level of joy in any room she was in. I hadn't met her father, but knew he was a journalist, as she was. My dad shared the news, and we got quiet. In the pause, the silence felt like a bowl for what was being felt. Sadness, of course, the collision of facts and disbelief, an ache at the thought of a friend facing a changed world with someone gone, but also an appreciation of my luck, too, the recognition that here we still were right now, my father and I.

He made his way back to his office, and I finished packing the tools, and pulled on a coat to head out for a walk. I passed by my dad, his back to me at his desk as he looked out the window at his birds. Now and then, the bullshit gets stripped away, and the accumulated anger and hurt and confusion give way for a glimpse at a different truth. And what I saw was that he was trying his best like all of us, eager and excited to share his enthusiasms about birds and fish and books, keeping the feeders stocked with sunflower seeds, fumbling like all of us to bring himself and his distracted love into focus. I was overwhelmed by a moment of crushing affection. Our friend's father would be dead soon. My dad had looked so happy when he saw the cases in their place.

"Bye, Dad," I yelled as I opened the door to head out for the walk, and my voice almost cracked.

Finally, the last pieces of trim went on, nail holes were filled, spots of paint touched up, and the cases were done. I grabbed a broom and swept. I scraped a bit of paint up off the floor. I put away the tools. The rug was rolled back, the big chair positioned again by the fireplace. The lamp went back to its home by the window, the bits of trash and scrap wood were deposited in a bin out the side door. I grabbed a beer and sat on the window seat facing the cases.

Dizzy from a long day and a fast beer, I thought again of the transformation that had just taken place. From soil and seed to big live tree with gnarled bark, from sawmill to board, from pieces one by one put together, sanded smooth, to this object, real in the world. First one thing, then something else. I made a nod of thanks to the trees who'd given their lives for these shelves, the same way my father would say some sort of non-god prayer of thanks when he caught a fish—not in gratitude at the catch, but to the fish for giving its life. A flush of gratitude toward trees rose in me. "Thanks, trees," I said out loud. One thing, then something else. I am not at home with change, suffer transitions the way most of us do. They're difficult, I think, because in quiet ways, transitions remind us of the final one.

My father was in the kitchen with his girlfriend. I called them in, and Jonah joined us, too. We sat, the four of us, on the window seat, which was too small for all of us, but we squeezed in and there were toasts and congratulations, exclamations. My father liked the shadows cast by the upper trim; his girlfriend noted how the top edge of one of the shelves lined up with the top edge of the thick wood mantel (a happy accident, as Mary would call it: total luck). I liked the seam between the case and the window trim. We clinked glasses and looked at the shelves and the shelves seemed to beam back at us.

"To books and bookshelves, creativity, hard work, and to family," my father said, raising his glass.

The next morning, we packed the car and said goodbye to the birds. A note came in from my brother Sam. "Heard you just finished shelves for Dad. Want to build me a big table? For my birthday?" I told him I'd love to. I took a photo of the shelves and sent it to Mary.

As we were walking out the door, I noticed the level had been left under the big chair. I placed it on a shelf and watched the bubble shift and settle. It wasn't right there in the middle. But it was good enough. No books would tumble off these shelves. I waved and honked and we headed home. As we pulled away, I was glad to know that I would be back again to see the shelves filled with books, to sit again in front of what I'd built, two bookcases to flank the fireplace.

Back in Cambridge, I called Mary and asked when I could drop off the tools I'd borrowed.

"Back door's unlocked," she said. "Come on up."

I drove over to Somerville, opened the gate to her backyard.

She'd pulled a tarp over the trash pile to keep the snow out. The heap had gotten big again, and I marveled at how fast we were able to produce so much, about how much had to be taken apart to be put back together again. I noticed a few jagged fragments of tub that hadn't been there before, chunks large and small, one, with clawfoot intact, that served to weight down the tarp.

I climbed the back stairs and came in through the kitchen.

"Is that your tub out there?" I asked.

"Come take a look."

I followed Mary to the bathroom. "It took about fifty whacks with the sledgehammer before the first crack. That thing was a beast."

"Mary, this looks amazing," I said of her renovation. It looked like guts, but it was the moment just before it all came together. She'd pulled up the old oak floor, taken down the walls, gotten rid of the sink, the tub, the toilet. Tufts of blown-in insulation drifted around. It was studs, joists, and pipes. "Check this out," she said, and I stepped in and she showed me how she'd framed for a sliding pocket door with a shelf on top, "for plants maybe, I don't know. New tub will go there." She pointed to the corner below the windows. "Toilet over there. And I just finished framing the shower today."

She explained that she'd just started adding new joists under the floor, to reinforce the area under where the tub would go, to support the extra weight. A couple fresh two-by-tens were nailed up tight against the dark old wood below the floor.

"Subway tiles?"

"I've subway-tiled enough bathrooms. I just couldn't handle it." She picked up a cream-colored four-by-four tile. "For the walls," she said. "Actually, I'm glad you reminded me."

I followed her out to the hall and she leaned over two boxes of twelve-by-twelve tiles. She pulled a few from each box and laid them on the floor. "Which do you like better?"

One set was a flat icy gray, lifeless and cold. It did not invite bare, post-bath feet. The others were light brown, sand colored, like the ones from the architect's bathroom from my first day on the job, with streaks of white and dark flecks, each one a little different and much warmer than the other option.

"These guys, definitely."

"Yeah, that's the way I was leaning."

"Is Emily psyched?" I asked, knowing she'd been pleading for this new bathroom for years now.

"She can't see it yet." I knew what she meant. When Emily looked in, she'd see chaos, mess, splintery wood, no walls and no floors, and pipes and wires. It would be hard to imagine how it would all become a room again. I could see it, in this moment before it all came together. I could picture just what it would look like, and moved through the steps quickly in my mind, layer over layer, until it looked like a bathroom.

I thanked her for the use of her tools, and for lending on-call support when I ran into trouble with the shelves. She waved her hand. "Of course."

"Really, Mary, thank you." And I hoped she understood that I wasn't just thanking her for the tools.

"Anytime you want to be the boss, go for it."

When Jonah came down to my dad's to lend a hand with the shelves, I'd been captain, showing him how to use the nail gun, pry trim off a window, use the level. He learned fast and we made a good, efficient team, a welcome surprise to learn that we could collaborate this way. In teaching I came to know I *knew* this stuff. But the thought of telling Mary what to do? No way.

"Give me a call if you want help tiling," I said.

As I was heading out, I ran into Emily in the hall.

She leaned in and lowered her voice. "She loved helping you with those shelves."

I blushed.

"No, I really mean it. She really loved it. Every time you asked a question. It made her so happy."

The clench in my throat and hot feeling in my face came as a surprise. I tried to swallow it down, didn't want her to see actual tears, and I thanked Emily again and told her how I loved having the help, that I couldn't have done it without Mary.

She leaned in closer and said, "Today's her birthday."

I went back to the bathroom and poked my head in the door. Mary was sitting on the edge of a sheet of plywood, her legs dangling down into the space below the floor. In her wool cap, she looked like a little kid.

"Happy birthday," I whispered.

She turned and smiled and shook her head.

"She tell you?"

I nodded.

"Get outta here, I gotta finish this up so we can go out to dinner."

We smiled at each other and I saw myself out. The blue painters' tape was still on the ceiling in the living room. The hole where the chimney used to be, still patched with a piece of plywood. I climbed down the back stairs and plaster crumbs crunched below my feet. Outside, the trash pile was a comfort, the wood and metal, the cracked tiles, swatches of insulation, door hinges, sawdust, dirt, all wrapped under big blue tarps, with fragments from her cast-iron tub holding it all in place. I pictured the demo guys coming back in the spring to load it all into the back of their big truck, giving Mary a hard time for letting some of the bags fill with water. And as soon as they roared off to bury it all in some trash grave, the pile would start to grow again, one bag and one board at a time.

The jobs change. We go in and out of other's people's homes. A room becomes a different room, altered, with some of its essence intact. Loose tiles become a floor. Boards become shelves. Wood becomes a wall. Places change. Homes change. Weather changes. We change.

How do we decide what's right for our own lives? The question never gets easier to answer. If we're lucky and we pay attention, pieces here and there will start to fit together. Parts shift into place, feel flush underneath the skin of the fingertip. For a moment, the bubble dips and shifts to show you level, at home with what you are, what you have become, and what you are becoming.

I stood on Mary's driveway and looked up the side of the house to the bathroom window on the second floor. The light blinked on and a moment later, the sound of hammerbangs rang out, new bones added to old skeleton, Mary upstairs on the floor with her hammer.

I got into the car and could hear the bangs. I rolled down the window, despite the cold, to hear them as I drove down her block. I stopped at the corner by the big brick church. The sound traveled through the evening air, echoed forward and back. Three more. *Bang, bang, bang.* Nail through wood. The light turned green and I made the turn toward home, heard one more before the roar of a city bus was the only thing I could hear. The hammerbangs echoed after that anyway. I heard the sound all the way home.

EPILOGUE

It's spring in Cambridge, and Mary and I began our fifth year working together last week, slamming new oak floors into a tiny office in the home of a retired sociology professor. It was, as it is every year, good to be back to it. After some months laboring on this book, it felt especially good to leave the inside of my apartment and the inside of my skull, and nail oak boards to a floor, feel the weight of the miter saw in my arms as I loaded it into Mary's van, and refamiliarize myself with the power of the crowbar. I felt the work in my shoulders and in my hamstrings when I woke up in the mornings. I walked home from work hungry and tired, happy to see the colorful rockets of crocus pressing out of the mulch in the still-light early evenings.

In the off-season, I made some tables, large ones and small. With each, I improved. (I cringe a bit to think of the first one I made, that long table for my brother as a birthday gift, functional but crude.) With each, I learned something I didn't know before. But more than the satisfaction of improvement, building them made me know how much there is to learn. Few things make one more aware of time, and time left, than facing all there is left to learn in a fresh endeavor. The tables stand. They are handsome and do their job as tables should. But there is so much I don't know.

Five years with Mary, and the work still feels new. In part, I think, it's exactly because of all that's left to learn. The poet Jon Cotner pointed me to the Korean proverb "Knows the way, stops seeing." It's not an argument for getting oneself lost, I think, but a nudge to stay awake, stay focused, alert even when time and experience have dulled us. Excitement arises from not knowing, from continuing to see, from trying to puzzle through. It's daunting—how much there is to know—and motivating, too. For now, I am satisfied to continue to get it wrong and try again and again to get it better and get it right.

With each table, with each wall and floor, with each set of bookcases built and filled so the shelves are heavy with books, comes the knowledge that all of it will fall apart someday. In our lifetimes, or after, these walls and floors and shelves will no longer do their jobs. The wood will splinter, rot, maybe get used for firewood, maybe get traded for some new model or discarded at the dump, scrap wood, sawdust, dead. This is their fate, and ours, as use and time enact their wearing. And I get carried away sometimes—I put my palm down on a sanded-smooth plank of black walnut and I feel the vibration there, I see the galaxy-swirl grain, and I think of the tree that stood to make this board, its roots veining deep into the soil, dark and warm, its long branches, its feather-shaped leaves in the wind, at sway and still. And here it is now, beneath my hand, joined, glued, clamped, turning into a table. *A thing's becoming other than it was.*

"What was is now no more," writes Ovid, "and what was not has come to be. Renewal is the lot of time." We are all unfinished after all.

Mary and I start a kitchen next week, right back into the thick of it. It will be hard work. "Do some push-ups this weekend," Mary told me as we parted ways on Friday afternoon. So I did.

ACKNOWLEDGMENTS

M att Weiland, the editor of this book, has the amazing ability to make *you have so much work to do* sound like *you can do it*. I'm so grateful I was able to work with someone as patient, funny, and wise. This book would not be, and would not be what it is, without him. My curious, spirited, and enthusiastic agent, Gillian MacKenzie, has been a candid source of guidance and support from the start, and I feel hugely lucky to be teamed up with her. I appreciate Nancy Green's sharp and sensitive copyediting, and am grateful to the efforts and hours the folks at Norton put in to help bring this book into being. And my thanks go to artist Joe McVetty for the elegant drawings of tools. I am inarticulatably grateful to my parents. To my mother, for the scarves, blankets, and mittens, for the wristlettes, hats, and turtle wraps, which is to say, for all the warmth. To my father, for his high standards. I'm especially grateful to my brothers, the two people I laugh the hardest with: to Will, who is the best storyteller I know, and to Sam, my trusted ally, to whom I turned more than anyone for help and feedback on this book. Pamela Murray's generosity, attention, and eagerness to celebrate occasions big and small has been a source of strength and warmth. I thank Goody-Goody, for the education. I thank Jenny White, for her kindness and care, her pep talks and perspective. I'll never know a better listener. I thank Alicia Simoni, whose depth of insight and understanding has helped me with this book and with much more than this book. I feel lucky to have shared meals and laughs with Joe and Laila Fontela. I'm grateful to Grub Street for helping me start, and to the *Boston Phoenix*, for being a wonderful way to spend my twenties. I give thanks to Richard Baker and Leona Cottrell, to Philip Connors, and to Mary, of course. And above all, to Jonah James Fontela, for whom love is an inadequate word.